Environmental Science and Engineering

The ultimate goal of this series is to contribute to the protection of our environment, which calls for both profound research and the ongoing development of solutions and measurements by experts in the field. Accordingly, the series promotes not only a deeper understanding of environmental processes and the evaluation of management strategies, but also design and technology aimed at improving environmental quality. Books focusing on the former are published in the subseries Environmental Science, those focusing on the latter in the subseries Environmental Engineering.

Shirsendu Banerjee · Sankha Chakrabortty ·
Jayato Nayak · Suraj K. Tripahy · Maulin Shah
Editors

Environmental Friendly Green Technologies for Improvement of Heavy Crude Oil Flow Assurance

Springer

Editors
Shirsendu Banerjee
School of Chemical Engineering
KIIT Deemed to be University
Bhubaneswar, Odisha, India

Sankha Chakrabortty
School of Chemical Engineering
KIIT Deemed to be University
Bhubaneswar, Odisha, India

Jayato Nayak
Centre for Life Science
Mahindra University
Hyderabad, Telangana, India

Suraj K. Tripahy
School of Chemical Engineering
KIIT Deemed to be University
Bhubaneswar, Odisha, India

Maulin Shah
Division of Applied and Environmental Microbiology
Enviro Technical Limited
Ankleshwar, Gujarat, India

ISSN 1863-5520 ISSN 1863-5539 (electronic)
Environmental Science and Engineering
ISBN 978-3-031-86700-2 ISBN 978-3-031-86701-9 (eBook)
https://doi.org/10.1007/978-3-031-86701-9

This Springer imprint is published by the registered company Springer Nature Switzerland AG
The registered company address is: Gewerbestrasse 11, 6330 Cham, Switzerland

Contents

Chapter 1
The Chemistry of Wax Formation, Agglomeration, and Deposition in Crude Oils

Biswadeep Pal, Sampa Guin, and Tarun Kumar Naiya

Abstract This chapter explores the complex mechanisms of wax-related phenomena in crude oil systems. Wax is a prevalent organic molecule that is commonly found in crude oils. Its existence presents considerable difficulties throughout the operations of extraction, transportation, and refining. This chapter provides a thorough examination of the fundamental chemistry involved in wax production, aggregation, and deposition. It aims to elucidate the intricate mechanisms responsible for these intricate occurrences. The chapter initially present a comprehensive overview of the molecular structures and properties of waxes, establishing a strong basis for comprehending their behaviour inside crude oils. The discussion also covered the temperature-dependent phase transitions of waxes, focusing on the important parameters that influence their crystallisation and deposition. Furthermore, the chapter explores the influence of operational circumstances, such as temperature and pressure, on the creation and stability of wax. Comprehending the aggregation of wax crystals is essential for understanding their behaviour during flow and transportation. The authors examine the influence of flow dynamics, shear stress, and pipeline geometry on wax deposition, providing useful insights for the development of effective mitigation techniques. Moreover, the chapter delves into the impact of several chemical additives and inhibitors in managing problems associated to wax. The text explores the methods by which these additions alter the process of wax crystallisation and impede deposition, hence improving the flow assurance characteristics of crude oils. This chapter offers a thorough examination of the scientific concepts that control the difficulties linked to wax in the petroleum sector. The information and ideas discussed in this chapter are essential for researchers, engineers, and practitioners who are aiming to improve the efficiency of crude oil production,

---

B. Pal · S. Guin · T. K. Naiya (✉)
Department of Petroleum Engineering, IIT (ISM) Dhanbad, Dhanbad, India
e-mail: trn2711@yahoo.com

B. Pal
e-mail: pal.biswadeep90@gmail.com

S. Guin
e-mail: sampaguin0912@gmail.com

S. Banerjee et al. (eds.), *Environmental Friendly Green Technologies for Improvement of Heavy Crude Oil Flow Assurance*, Environmental Science and Engineering,
https://doi.org/10.1007/978-3-031-86701-9_1

transportation, and refining. These discoveries also help ensure smooth flow and operational effectiveness.

Keywords Wax formation · Crude oil · Flow assurance · Pipeline transportation · Wax mitigation

1.1 Introduction

Crude oil is a complex organic mixture primarily composed of hydrocarbons, with some non-hydrocarbon compounds. Its composition is generally categorized into saturates, aromatics, resins, and asphaltenes (Xiong et al. 2020a, b). Certain saturated compounds have high melting points and solidify easily at lower temperatures, forming colloidal structures known as wax. This wax precipitates from the fluid when the temperature drops below approximately 80 °C. The wax-forming molecules mainly include normal and branched alkanes, although some naphthenes with long carbon chains can also contribute.

Wax deposition poses significant economic and operational challenges in the oil and gas industry, occurring during production, transportation, and storage. Common issues include blockages in well tubing and pipelines, reduced drawdown pressure, fouling of surface facilities like separators and tankers, and increased energy requirements for pumps. As global energy demand continues to rise exponentially, conventional oil reserves are rapidly depleting. This trend has driven the exploration of alternative energy sources and spurred production from offshore reserves and extreme climatic environments over the past decade (Bai and Bai 2018).

Transporting crude oil from such fields often requires long-distance pipelines. Without adequate control measures, the oil temperature can drop to environmental levels, triggering wax precipitation. Severe wax deposition can lead to significant operational disruptions, potentially jeopardizing the economic feasibility of projects due to the high costs of remediation. This issue is particularly challenging as numerous waxy oil fields exist worldwide, and no universal solution has been identified (Batsberg et al. 1991; Del Carmen Garcia 2001; Ding et al. 2006; Rønningsen 2012; Pedersen et al. 1991; Suppiah et al. 2010).

Many companies employ pigging as the preferred remediation technique. However, this method has drawbacks, including production interruptions and additional costs associated with its frequent implementation, such as energy consumption and maintenance expenses (Niesen 2002; Xie et al. 2018).

1.2 Structural Composition

Bishop et al.'s experimental study on high molecular weight hydrocarbons revealed that wax-forming alkanes primarily consist of lighter compounds with carbon numbers in the range of C20-C50. In contrast, heavier compounds with carbon numbers exceeding C50 are rarely found in solid wax (Philp et al. 1995). This finding is linked to the distribution pattern of alkanes/paraffins, with normal alkanes and slightly branched iso-paraffins present in significant quantities in the lighter C7+ fractions.

In the heavier fractions, the degree of molecular branching increases, reducing the likelihood of these molecules forming solid structures. While alkane components with minimal or no branching are abundant in the lighter C7+ fractions, their contribution to the solid wax phase is limited by their lower melting points. Consequently, the solid wax phase predominantly consists of alkanes in the C20–C50 range. The general composition of wax, which mainly includes normal and branched alkanes along with naphthenes featuring long alkane chains capable of precipitating, is illustrated in Fig. 1.1 (Kiyingi et al. 2022).

The waxy behavior of crude oil is primarily attributed to the presence of paraffin, which encompasses a mixture of n-alkanes and isoalkanes with carbon chains ranging from C20 to C50. Paraffin precipitates from crude oil when the temperature falls below the wax appearance temperature (WAT). This precipitation increases viscosity and reduces flowability during transportation.

Experimental studies have examined the relationship between paraffin content and the properties of crude oil. Factors such as paraffin crystallization, nanoparticle aggregation, and others are known to influence viscosity changes in oil and oil/water emulsions. However, the molecular-level mechanisms driving WAT and viscosity variations remain poorly understood.

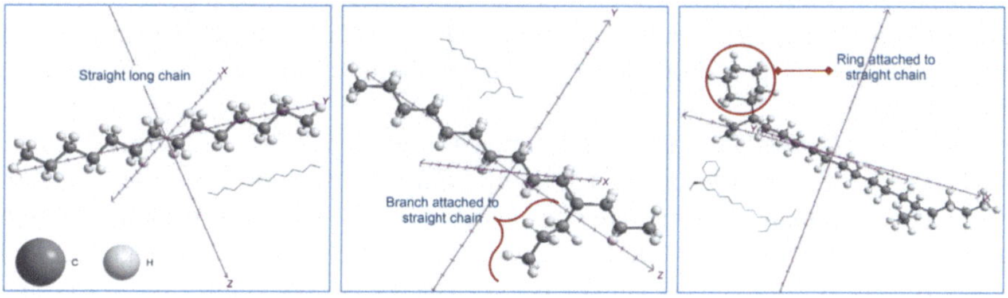

Fig. 1.1 Structural composition of waxes in crude oils

1.3 Wax Deposition Mechanism

The primary factor influencing the wax deposition process is low temperature, making subsea pipelines particularly susceptible. Consequently, preventing wax deposition is critical in deep-water oil production. Wax deposition in crude oil systems can be mitigated or prevented using chemical, mechanical, thermal, or a combination of these remediation methods. However, with the increasing prevalence of ultra-deepwater production, offshore drilling, and ocean-floor completions, mechanical and thermal methods often become economically unfeasible. As a result, the use of chemical additives as wax deposition inhibitors is becoming more widespread.

Thermal methods include techniques such as heat retention, active heating (e.g., thermal insulation), bottom-hole heaters, hot oil circulation, and steam circulation. Mechanical methods involve tools like scrapers in boreholes and pipeline pigging at regular intervals. Chemical inhibitors include pour point depressants (PPDs), crystal modifiers, dispersants, and solvents.

It is essential to distinguish between wax deposition and wax precipitation, as the two are often used interchangeably in literature despite their differences. Wax deposition refers to the formation and growth of a solid-phase layer on a surface in contact with crude oil, while wax precipitation pertains to the formation of solid wax particles in the bulk fluid. Wax deposition can occur through mechanisms such as shear dispersion, gravity settling, and Brownian motion from precipitated wax or molecular diffusion from dissolved wax. Near the cold pipe wall, precipitated wax molecules form an incipient gel—a 3D network of wax crystals trapping significant amounts of oil. This gel layer grows over time due to radial thermal and mass transfer gradients caused by heat loss to the surroundings, as illustrated in Fig. 1.2.

Several factors influence wax deposition in pipelines, including pipe wall temperature, crude oil composition, crude oil temperature, ambient temperature, flow rate, thermal history, time, and pressure. Wax deposition increases with a greater temperature difference between the bulk wax solution and the pipe's cold surface. Deposition occurs when the surface temperature is below both the solution temperature and the cloud point.

Crude oil composition significantly impacts wax deposition and affects properties like pour point and viscosity. Crude oil comprises various molecules, ranging from light components like methane, responsible for solid hydrate formation at high pressure and low temperature, to heavy components such as long-chain alkanes and isoparaffins, which tend to crystallize into solid phases at low temperatures, forming both macro- and microcrystalline structures.

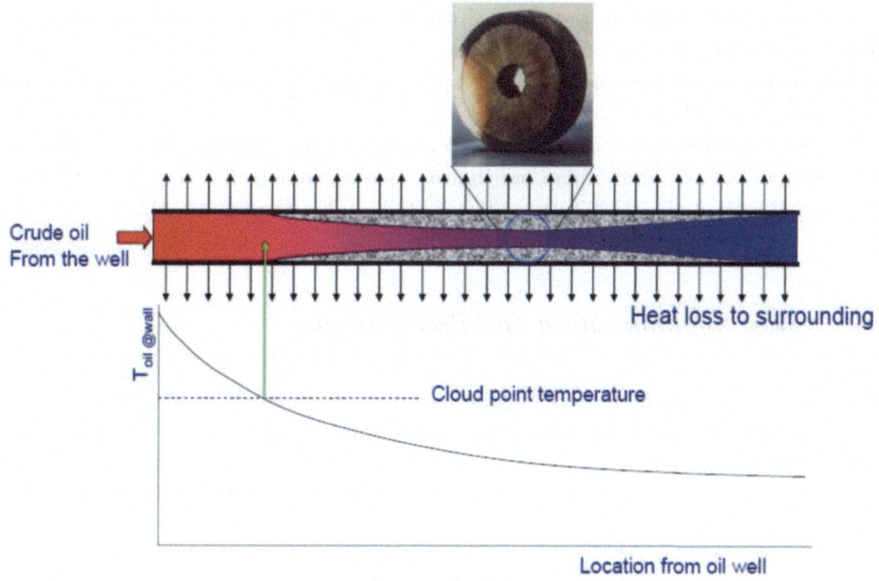

Fig. 1.2 Relation between wax deposition and cloud point (Al-Yaari 2011)

1.3.1 Wax Formation

As oil temperatures decrease beyond a certain threshold, wax may begin to form a solid layer inside the pipeline. If left unaddressed, this layer can accumulate during continued transport, eventually leading to pipeline blockages. However, not all wax precipitates deposit on the pipeline walls; some crystals remain suspended as solid particles within the liquid phase (Fig. 1.3).

This process is primarily influenced by temperature and crude oil composition. When the oil temperature falls below the wax appearance temperature (WAT),

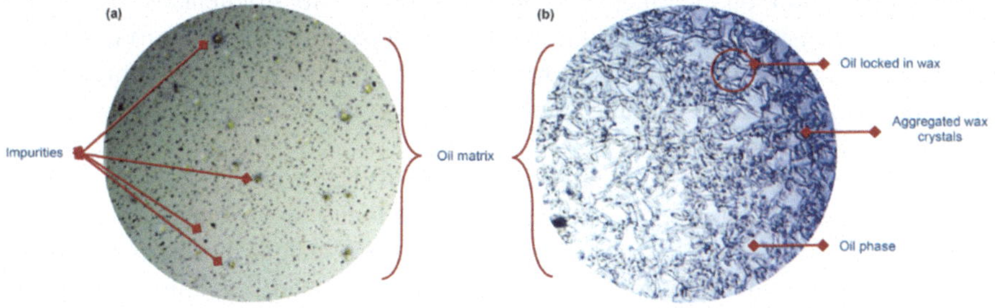

Fig. 1.3 Microscopic view of wax crystallization in oil using commercial wax (high power magnification 40) **a** Blank oil sample; **b** Oil mixed with 2 wt% commercial wax (Kiyingi et al. 2022)

paraffin molecules begin to precipitate from the liquid phase, forming small crystalline solids. With further temperature drops, these solids can aggregate, creating a gel-like matrix consisting of trapped oil and wax particles, which then deposit onto the pipeline walls (Fig. 1.3b).

While temperature and oil composition are key factors affecting wax deposition, research has also explored other contributing factors, such as flow regimes and flow velocities (Quan et al. 2020; Taheri-Shakib et al. 2020; Zhou et al. 2020).

1.3.2 Wax Agglomeration and Precipitation

Initially, mechanisms such as Brownian motion, molecular diffusion, shear dispersion, and gravity were believed to drive wax formation. However, over time, some of these theories have been challenged, and additional mechanisms, including nucleation, shear stripping, and aging, have been proposed (Haj-Shafiei et al. 2019; Hoffmann et al. 2012).

1.3.2.1 Brownian Motion, Shear Dispersion, and Gravity

Some studies have questioned the validity of certain wax deposition mechanisms for various reasons. For instance, Brownian motion—the random movement of small particles—was initially proposed as a cause of wax deposition, suggesting that wax molecules could accumulate at the center of pipelines (Burger et al. 1981). However, this mechanism has been criticized because the lower temperature at the pipeline walls compared to the system temperature makes it unrealistic to attribute wall deposition to Brownian motion (Singh et al. 2000a, b).

Similarly, gravity-driven effects, attributed to differences in the densities of oil and wax particles, suggest that wax particles settle due to gravitational forces. While this mechanism may be applicable to horizontal pipelines, it is less plausible in vertical tubes or pipes (Burger et al. 1981; Mansourpoor et al. 2019).

Shear dispersion, on the other hand, arises from velocity gradients within the oil flow (Hsu et al., 1994). As crude oil flows through a pipeline, the shear forces acting on each unit area contribute to the movement of the medium. This shearing energy drives wax crystal particles in the crude oil to migrate, accumulate, and grow, as illustrated in Fig. 1.4.

It is suggested that the velocity gradient, caused by the decreasing particle velocity near the pipe walls compared to the maximum velocity at the center of the flow, contributes to wax deposition on the pipe walls. While this mechanism appears plausible under laminar flow conditions, it does not adequately explain deposition in more dynamic scenarios, such as turbulent flow, where particles are more likely to be dislodged and carried into the bulk fluid (Bird 2002; Lee et al. 2020; Saffman 1965). Consequently, this mechanism has been largely disregarded in many wax deposition models.

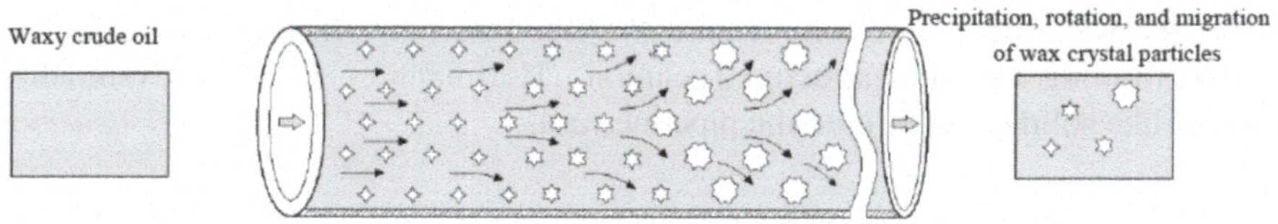

Fig. 1.4 Mechanism of shear dispersion during crude oil flow

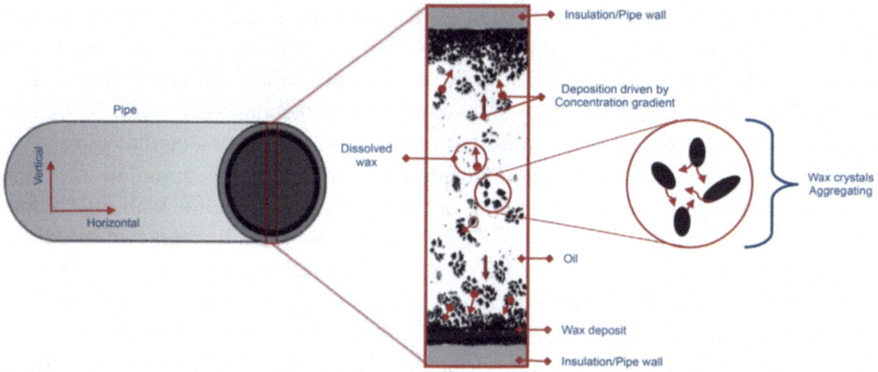

Fig. 1.5 Molecular diffusion mechanism during crude oil flow, Kiyingi et al. (2022) (Open access)

1.3.2.2 Molecular Diffusion

Molecular diffusion, as illustrated in Fig. 1.5, describes the movement of wax molecules from regions of higher concentration (center of the pipe) to regions of lower concentration near the pipe walls, driven by a temperature gradient as the oil temperature approaches and drops below the wax appearance temperature (Burger et al. 1981; Hoffmann et al. 2012; Matzain 1996; Singh et al. 2000a, b). This mechanism is now widely recognized within the scientific community as the primary driver of wax deposition (Alnaimat and Ziauddin 2020; Huang et al. 2011a,b; Wang et al., 2013). It is commonly incorporated into depositional models using Fick's law of diffusion.

1.3.2.3 Nucleation/Emulsified Nucleation

This mechanism has garnered significant attention due to the formation of emulsions during oil production, which can occur for various reasons, including pump actions. Under low-temperature conditions, emulsified particles coalesce, creating a surface for wax molecules to deposit. Wax formation begins with the nucleation of minute solid particles, followed by their growth as additional molecules attach. During this process, the tiny crystals adsorb onto the dispersed phase, leading to agglomeration and the development of a wax crystal network as the temperature continues to drop below the wax appearance temperature (WAT). This network traps oil within its matrix, forming a gel-like structure (gelation) (Hoffmann et al. 2012). Eventually, this gel-like solid deposits onto the pipeline walls.

1.3.2.4 Aging

Wax molecules on the surface of wax deposits infiltrate deeper into the deposit, strengthening its mechanical properties, such as thickness and hardness (Hernandez et al. 2003; Huang et al. 2011a,b; Matzain 1996). Some studies have specifically focused on the changes that occur in wax deposits after their initial formation. This is crucial because, in addition to understanding the mechanisms involved in wax deposition in pipelines, it is essential to comprehend the processes that govern the aging of wax deposits to fully model crude oil flow and wax accumulation. These deposits are not static; once a layer of wax forms along the pipeline wall, its composition gradually evolves. The crystalline wax deposit behaves like a porous medium, trapping oil within its three-dimensional network (Singh et al. 2000a, b, 2001). As time progresses, the wax content in the gel deposit can increase by diffusion. This process affects properties such as hardness, melting point, and heat of fusion, which in turn influence the selection of appropriate wax removal methods in pipelines. Singh et al. (2000a, b) identified a counter-diffusion phenomenon, where wax molecules diffuse into the gel deposit and oil molecules diffuse out, as the key factor in the aging of wax deposits. They also found that the aging rate depends on oil flow rate and pipeline wall temperature. In their experimental setup, where oil flowed in a closed loop with cooled walls, there was a rapid decrease in the internal radius during the first day, followed by a plateau. Similarly, the increase in the wax content (measured by weight fraction) slowed after the first day's rapid change. The wax content of the gel deposit, as determined by high-temperature gas chromatography (HTGC), changes over time, with the proportion of lighter components decreasing and the proportion of heavier components increasing after the first day. Singh et al. found that wax content continued to rise even after the thickness of the deposit stabilized. Wax molecules with chain lengths greater than 29 carbon atoms diffused into the deposit, while those with shorter chain lengths diffused out. The value of 29 carbon atoms is considered the critical carbon number (CCN) for the given operational conditions and can help determine the appropriate inhibitors to use, based on whether they can prevent crystallization of waxes above the CCN (Paso and Fogler 2003). This mechanism is influenced by temperature gradients and varies with different crude oil compositions. It is reasonable to suggest that shear stripping and this diffusion mechanism complement each other, as effused oil molecules are stripped away while wax fills the space left behind.

1.3.2.5 Shear Stripping

The removal of wax deposits through shear forces becomes particularly crucial under turbulent flow conditions, as the rate of removal is significantly higher compared to laminar flow. Therefore, to accurately model wax deposition, especially in turbulent flow, it is essential to include shear stripping effects in the model. Furthermore, incorporating shear force-based wax removal into the model could assist in the design of flow improver chemicals, as some of these may work by promoting the formation of softer gel structures that are more easily removed by shear forces. Researchers, such as Matzain (1999), have attempted to represent this effect through empirical correlations to quantify the reduction in the rate of deposit formation caused by shear forces.

1.3.2.6 Nucleation and Gelation Kinetics

The crystallization of waxes is a kinetic process, which can be described by classical homogeneous nucleation theory (Paso 2005). Although much research has been devoted to approaching wax deposition as a thermodynamic issue, there has been limited exploration of modeling wax deposition based on the kinetics of deposit formation (Merino-Garcia et al., 2007). Paso (2005) aimed to address the insufficient understanding of the crystallization and gelation processes, as well as the assumption that paraffin precipitation kinetics do not limit deposition rates—an assumption that could lead to inaccurate predictions of wax deposition in situations where a stable gel cannot form. Using model fluids made of n-paraffin components dissolved in petroleum mineral oils, Paso applied homogeneous nucleation and crystallization theory along with differential scanning calorimetry to measure the onset of crystallization and the crystallization rate.

Paso (2005) compared experimental and equilibrium crystallization rates, identifying three regimes in the crystallization process at low cooling rates. The first regime is a nucleation lag period, starting at high-temperature conditions. The second regime is a supersaturation growth period, driven by the supersaturation established during the nucleation lag period and decreasing solubility conditions, during which the crystallization rate can significantly exceed the equilibrium rate. The third regime is an equilibrium growth period, which begins when the supersaturation ratio diminishes, and the crystallization rate aligns with the equilibrium predictions based on the van't Hoff relation. Paso observed that the temperature span of the supersaturation growth regime was independent of the model fluid viscosity, indicating no transport limitations in the crystallization rate.

By applying the van't Hoff solubility model within the context of classical homogeneous nucleation theory, Paso (2005) demonstrated that nucleation is the primary kinetic limitation in the crystallization of n-alkanes in organic solutions under low cooling rate conditions, with crystallization rate limitations becoming more significant at higher cooling rates. He also noted that the initial nucleation event depends on the solubility behavior of the highest fraction of n-alkane components in the fluid, and that introducing chain-length variations reduces the critical nucleus surface energy through co-crystallization of paraffins with different chain lengths.

Additionally, Paso (2005) investigated the mechanical properties of waxy model fluids under constant cooling rates using controlled-stress rheometric measurements. By applying oscillatory stress to fluid samples, he characterized their mechanical properties during gelation. Microscopic analysis was used to study the crystal structure in the samples, and Paso extended an established three-dimensional analytical percolation model to wax–oil gel systems. This allowed for the theoretical prediction of gelation via the percolation threshold—the fractional volume of the solid crystalline phase at which it forms a continuous, domain-spanning path linked by crystal–crystal interactions. In this context, paraffin crystals were represented as ellipsoidal geometries with spherical rotational volumes of interaction. The primary and secondary ellipsoidal aspect ratios of the crystals, α_1 and α_2, were related to the solid phase fraction at the percolation threshold, Øg, by a specific equation.

$$Ø_g = \theta_p\left(\frac{1}{\alpha_1}\frac{1}{\alpha_2}\right) \tag{1.1}$$

Here $\theta_p = 0.295 =$ represents the spherical percolation threshold. While this model predicts the formation of a crystal percolation network, it was observed that gelation will only occur if the number density and strength of the crystal–crystal interactions are sufficient to impart solid-like properties to the fluid. Paso (2005) highlighted that the gel point of a waxy petroleum fluid depends on the morphologies and surface characteristics of the randomly oriented paraffin crystals. He found that aspect ratios on the order of 100 allowed mechanical gels to form from oils with paraffin content as low as 0.5%. Additionally, he noted that monodisperse crystals, with ordered surfaces and sharp edges, had minimal crystal–crystal contact and weak interactions, whereas polydisperse n-alkane crystals, with nanoscale surface roughness, provided contact points for stronger crystal–crystal interactions, enabling mechanical gelation at lower wax concentrations. Paso concluded that percolation threshold models accurately predict the gel point for physical gelation systems with strong crystal–crystal interactions but tend to under-predict the solid fraction required to induce gelation in systems with weakly interacting particles.

1.4 Inhibition of Wax Deposition

The most effective approach to addressing wax deposition in crude oil pipelines is to prevent it from occurring initially. As a result, researchers have explored various methods to inhibit the deposition process. These include heat insulating subsea pipelines to maintain higher temperatures and prevent precipitation (Quenelle and Gunaltun 1987), applying internal coatings of plastics to pipeline surfaces (Patton 1970; Bummer 1971), and using chemical inhibitors to prevent wax deposition on the walls of the pipeline.

1.4.1 Chemical Inhibitors

Numerous researchers have investigated the effectiveness of various wax deposition inhibitors and the mechanisms by which they prevent deposition. Notable studies include those by Jorda (1966), Mendell and Jessen (1970), Fulford (1975), Addison (1984), Newberry and Barker (1985), Fielder and Johnson (1986), Singhal et al. (1991), Jang et al. (2007), and Tinsley et al. (2007). For example, Wang et al. (2003) found that while the inhibitors they tested reduced the overall amount of deposition, they were only partially successful in preventing the deposition of high molecular weight paraffin components (C35 and above), leading to harder wax deposits compared to untreated oil. They also discovered that inhibitors most effective at lowering the wax appearance temperature (WAT) were better at reducing total wax deposition, and that adding the corrosion inhibitor oleic imidazaline (OI) significantly improved the effectiveness of the wax deposition inhibitors. Figure 1.6 shows their results, where PIE represents the paraffin inhibition efficiency, indicating the amount of wax deposited with the inhibitor as a percentage of the amount deposited without it. Bello et al. (2006) also examined the performance of commercial wax inhibitors, particularly with Nigerian crude oils, and found that a trichloroethylene–xylene (TEX) binary system was more effective and economically viable than commercial inhibitors. Other researchers have emphasized the importance of tailoring inhibitor treatments to specific crude oils for maximum efficacy. For instance, Manka and Ziegler (2001) showed that additives perform best when matched to the paraffin distribution in the crude being treated. Similarly, Carmen García (2001) highlighted the strong correlation between the efficiency of a specific paraffin inhibitor and the composition of the crude oil, which necessitates case-by-case selection of inhibitors for field applications.

The effectiveness of commercially available wax inhibitors is often limited and must be assessed on a case-by-case basis. For example, Wang et al. (2003) found that while the wax inhibitors they tested reduced overall deposition, they had limited success in preventing the deposition of high molecular weight paraffin components (C35 and above). As a result, the wax deposits formed were harder compared to untreated oil. They also discovered that inhibitors that were most effective at lowering the wax appearance temperature (WAT) were more successful in reducing total wax deposition. Additionally, the inclusion of the corrosion inhibitor oleic imidazaline (OI) significantly enhanced the effectiveness of the deposition inhibitors. Figure 1.6 illustrates some of their findings, where PIE represents the paraffin inhibition efficiency, indicating the amount of wax deposited with the inhibitor as a percentage of the amount deposited without it. Bello et al. (2006) also assessed the effectiveness of commercial wax inhibitors, specifically with Nigerian crude oils, and found that a trichloroethylene–xylene (TEX) binary system was more effective and economically viable than conventional commercial inhibitors. Other studies have highlighted the importance of tailoring inhibitor treatments to the specific characteristics of the crude oil to optimize effectiveness. For instance, Manka and Ziegler (2001) observed that additives perform best when aligned with the paraffin distribution of the crude

Inhibitor	PIE_{light} (%)	PIE_{heavy} (%)	PIE_{total} (%)
A	69	-15	41
A_OI	70	-4	46
B	89	68	82
B_OI	92	79	88

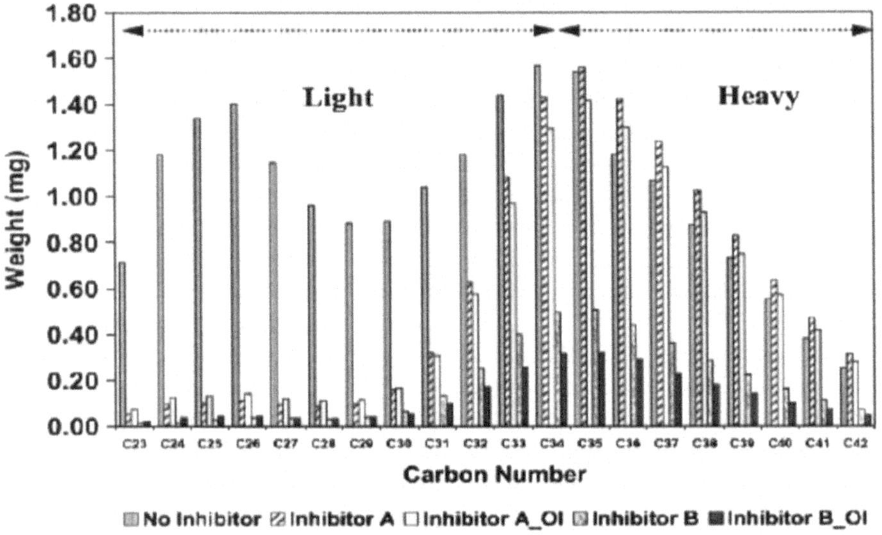

Fig. 1.6 Effect of wax inhibitors (100 ppm) and oleic imidazoline, OI, (200 ppm) on paraffin deposition from a mixture of paraffin wax in C10 solution (Wang et al. 2003)

oil being treated. Likewise, Carmen García (2001) pointed out a strong correlation between the efficiency of a specific paraffin inhibitor and the crude oil composition, suggesting that inhibitor selection requires careful consideration for each individual case. Additionally, environmental factors must be considered when choosing a wax inhibitor, especially in operations at low temperatures, as the inhibitor formulation must be winterized to ensure effective delivery under such conditions (Manka et al. 1999; Jennings and Breitigam 2009). While research into more effective wax inhibitors continues, it remains true that inhibitors typically do not achieve 100% inhibition. As such, they are often used in combination with other remediation methods, such as pigging (Jennings and Breitigam 2009; Kelland 2009).

1.4.2 Surfaces that Prevent Wax Deposition

Developing wax-repellent surfaces for oil pipelines holds significant appeal, as it could reduce or eliminate the need for wax inhibition and removal techniques to ensure smooth operation. By gaining a deeper understanding of the mechanisms by which wax adheres to pipeline walls, it may be possible to design pipelines

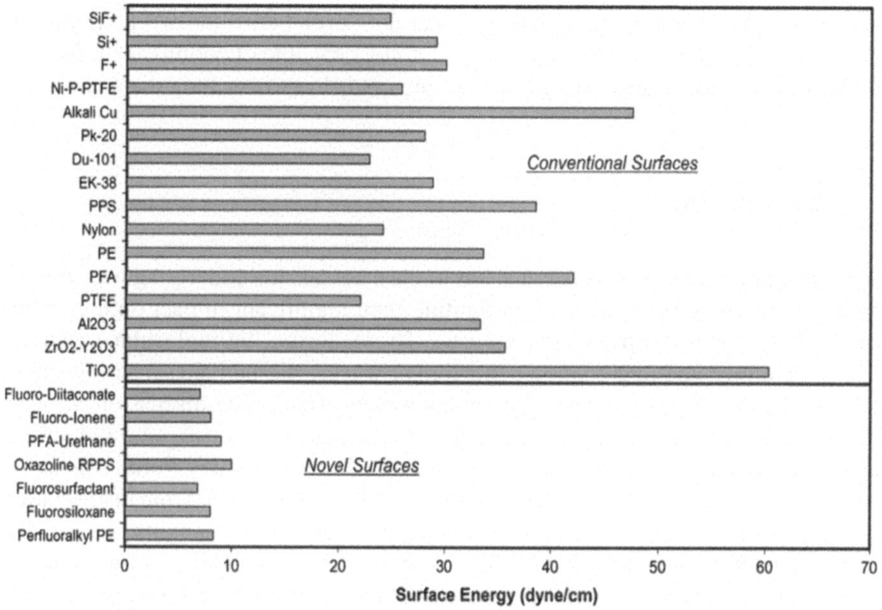

Fig. 1.7 Surface energy reduction possible with novel surface technologies (Paso et al. 2009)

where the wall properties make adhesion unfavorable. Paso et al. (2009) conducted a thorough review on the use of non-stick and anti-adhesive coatings to prevent solid–liquid deposition, exploring metal surface treatments and synthesized polymers. The materials they identified as promising included fluoro-siloxanes, fluoro-urethanes, oxazolane-based polymers, and hybrid diamond-like carbon and polymer coatings. Figure 1.7 displays the surface free energies of various surfaces tested for paraffin control, as reported by Paso et al. (2009), providing insight into the ability of these surfaces to interact with waxes and their potential to prevent wax deposition. Continued research into the mechanisms of wax adhesion is expected to lead to the development of even more effective surface treatments in the future.

1.4.3 Cold Flow

Heating or insulating subsea pipelines can help prevent the pipeline wall temperature from dropping below the WAT. However, an alternative method for inhibiting wax deposition, explored by Merino-Garcia and Correra (2008), is the use of Cold Flow technology. This approach proposes that wax deposition could be prevented by reducing the bulk temperature within the pipeline to match the surrounding seawater temperature, effectively eliminating the temperature gradient. This would allow the waxes to remain as a solid dispersion within the bulk fluid. While wax deposition may become negligible in the absence of heat flux, even below the WAT, significant development is still needed to create the technology capable of cooling the bulk fluid to this condition and transporting the resulting cold slurry over long distances.

1.5 Conclusion

This chapter provides an in-depth analysis of the mechanisms underlying wax-related challenges in crude oil systems, highlighting their significant impact on the operational efficiency of the petroleum industry. By exploring the molecular structure, phase transitions, and aggregation behaviors of waxes, it establishes a strong scientific foundation for addressing these complex phenomena. The discussion of operational parameters and the influence of flow dynamics on wax deposition identifies key factors affecting crude oil transportation and refining processes. Furthermore, the review of chemical additives and inhibitors emphasizes their crucial role in mitigating wax-related issues and enhancing flow assurance. Together, the insights presented in this chapter offer valuable guidance for advancing research and developing innovative strategies to optimize crude oil production, transportation, and processing. This understanding enables industry stakeholders to implement more effective solutions, ensuring the reliability and sustainability of operations within the petroleum sector.

References

Addison GE (1984) Paraffin control more cost-effective. In: SPE eastern regional meeting. Society of Petroleum Engineers, Charleston

Alnaimat F, Ziauddin M (2020) Wax deposition and prediction in petroleum pipelines. J Petrol Sci Eng 184:106385. https://doi.org/10.1016/j.petrol.2019.106385

Al-Yaari M (2011) Paraffin wax deposition: mitigation and removal techniques. Presented at the SPE Saudi Arabia section young professionas technical symposium, Dhahran, 14–18 Mar 2011 (SPE 155412 MS)

Bai Y, Bai Q (2018) Subsea engineering handbook. Gulf Professional Publishing

Batsberg PW, Hansen BA, Larsen E, Nielsen AB, Roenningsen HP (1991) Wax precipitation from North sea crude oils 2 solid-phase content as function of temperature determined by pulsed NMR. Energy Fuels 5(6):908e913. https://doi.org/10.1021/ef00030a020

Bello OO, Fasesan SO, Teodoriu C, Reinicke KM (2006) An evaluation of the performance of selected wax inhibitors on paraffin deposition of Nigerian crude oils. Pet Sci Technol 24:195–206

Bird RB (2002) Transport phenomena. Appl Mech Rev 55(1):R1–R4

Bummer BL (1971) Improved paraffin prevention techniques reduce operating costs, powder river basin, wyoming. In: Rocky mountain regional meeting of the society of petroleum engineers of AIME. American Institute of Mining, Metallurgical, and Petroleum Engineers, Billings

Burger E, Perkins T, Striegler J (1981) Studies of wax deposition in the trans Alaska pipeline. J Petrol Technol 33(6):1075–1086. https://doi.org/10.2118/8788-PA

Carmen García M (2001) Paraffin deposition in oil production. In: SPE international symposium on oilfield chemistry. Society of Petroleum Engineers, Houston

Del Carmen Garcia M (2001) Paraffin deposition in oil production. In: Paper presented at the SPE international symposium on oilfield chemistry. https://doi.org/10.2118/64992-MS

Ding J, Zhang J, Li H, Zhang F, Yang X (2006) Flow behavior of Daqing waxy crude oil under simulated pipelining conditions. Energy Fuels 20(6):2531e2536. https://doi.org/10.1021/ef060153t

Fielder M, Johnson RW (1986) The use of pour-point depressant additive in the beatrice field. In: SPE European petroleum conference. Society of Petroleum Engineers, London

Fulford RS (1975) Oilwell paraffin prevention chemicals. In: SPE regional meeting. American Institute of Mining, Metallurgical, and Petroleum Engineers, Oklahoma

Haj-Shafiei S, Workman B, Trifkovic M, Mehrotra AK (2019) In-situ monitoring of paraffin wax crystal formation and growth. Cryst Growth Des 19(5):2830e2837. https://doi.org/10.1021/acs.cgd.9b00052.

Hernandez O, Hensley H, Sarica C, Brill J, Volk M, Delle-Case E (2003) Improvements in single-phase paraffin deposition modeling. In: Paper presented at the SPE annual technical conference and exhibition. https://doi.org/10.2118/84502-PA

Hoffmann R, Amundsen L, Huang Z, Zheng S, Fogler HS (2012) Wax deposition in stratified oil/water flow. Energy Fuels 26(6):3416e3423. https://doi.org/10.1021/ef2018989

Hsu J, Santamaria M, Brubaker J (1994) Wax deposition of waxy live crudes under turbulent flow conditions. In: Paper presented at the SPE annual technical conference and exhibition. https://doi.org/10.2118/28480-MS

Huang Z, Zheng S, Fogler HS (2015) Wax deposition: experimental characterizations, theoretical modeling, and field practices. Taylor & Francis Group, USA

Huang Z, Lee HS, Senra M, Fogler HS (2011a) A fundamental model of wax deposition in subsea oil pipelines. AlChE J 57(11):2955e2964. https://doi.org/10.1002/aic.15750

Huang Z, Lu Y, Hoffmann R, Amundsen L, Fogler HS (2011b) The effect of operating temperatures on wax deposition. Energy Fuels 25(11):5180e5188. https://doi.org/10.1021/ef201048w

Jang YH, Blanco M, Creek J, Tang Y, Goddard WA (2007) Wax Inhibition by comb-like polymers: support of the incorporation-perturbation mechanism from molecular dynamics simulations. J Phys Chem B 111:13173–13179

Jennings DW, Breitigam J (2009) Paraffin inhibitor formulations for different application environments: from heated injection in the desert to extreme cold arctic temperatures. Energy Fuels 24:2337–2349

Jorda RM (1966) Paraffin deposition and prevention in oil wells. SPE J Petrol Technol 18:1605–1612

Kelland MA (2009) Production chemicals for the oil and gas industry. CRC Press

Kiyingi W, Guo J, Xiong R, Su L, Yang X, Zhang S (2022) Crude oil wax: a review on formation, experimentation, prediction, and remediation techniques. Pet Sci 19(5):2343–2357. https://doi.org/10.1016/j.petsci.2022.08.008

Lee DG, Lim JS, Kim YJ, Woo NS, Han SM, Ha J (2020) Monitoring and detection of paraffin wax deposition process based on ultrasonic analysis. J Nanosci Nanotechnol 20(1):168e176. https://doi.org/10.1166/jnn.2020.17282

Manka JS, Ziegler KL (2001) Factors affecting the performance of crude oil wax control additives. In: SPE production and operations symposium. Society of Petroleum Engineers, Oklahoma

Mansourpoor M, Azin R, Osfouri S, Izadpanah A (2019a) Experimental measurement and modeling study for estimation of wax disappearance temperature. J Dispersion Sci Technol 40(2):161e170. https://doi.org/10.1080/01932691.2018.14616-35

Matzain A (1996) Single phase liquid paraffin deposition modeling. University of Tulsa

Matzain A (1999) Multiphase flow paraffin deposition modeling. The University of Tulsa, Tulsa, Oklahoma

Mendell JL, Jessen FW (1970) Mechanism of inhibition of paraffin deposition in crude oil systems. In: Ninth biennial production techniques symposium. American Institute of Mining, Metallurgical and Petroleum Engineers, Wichita Falls

Merino-Garcia D, Correra S (2008) Cold flow: a review of a technology to avoid wax deposition. Pet Sci Technol 26:446–459

Merino-Garcia D, Margarone M, Correra S (2007) Kinetics of waxy gel formation from batch experiments. Energy Fuels 21:1287–1295

Newberry ME, Barker KM (1985) Formation damage prevention through the control of paraffin and asphaltene deposition. In: SPE production operations symposium. Society of Petroleum Engineers, Oklahoma City

Niesen V (2002) The real cost of subsea pigging. EP Mag:97

Paso KG (2005) Paraffin gelation kinetics (PhD Dissertation) University of Michigan Ann Arbor, Michigan

Paso KG, Fogler HS (2003) Influence of n-paraffin composition on the aging of wax-oil gel deposits. AIChE J 49:3241–3252

Paso KG, Kompalla T, Aske N, Rønningsen HP, Øye G, Sjöblom J (2009) Novel surfaces with applicability for preventing wax deposition: a review. J Dispersion Sci Technol 30:757–781

Patton CC (1970) Paraffin deposition from refined wax-solvent systems. Soc Petrol Eng J:17–24

Pedersen KS, Skovborg P, Roenningsen HP(1991) Wax precipitation from North Sea crude oils 4 thermodynamic modeling. Energy Fuels 5(6):924e932. https://doi.org/10.1021/ef00030a022

Philp RP, Bishop A, Del Rio JC, Allen J (1995) Characterization of high molecular weight hydrocarbons (>C40) in oils and reservoir rocks. Spec Publ Geol Soc Lond 86(1):71e85. https://doi.org/10.1144/GSL.SP.1995.086.01.06

Quan Q, Ran W, Wang S, Wang Y, Li R, Gong J (2019) Prediction the variation range of wax deposition with temperature of crude oil in a flow loop. Petrol Sci . Technol 37(15):1739e1746. https://doi.org/10.1080/10916466.2018.1511580

Quan Q, Wang S, Sun N, Wang Y, Li R, Gong J (2020) Experimental study on wax deposition of gas-liquid under intermittent flow. Petrol Sci Technol 38(4):331e337. https://doi.org/10.1080/10916466.2019.1705854

Quenelle A, Gunaltun M (1987) Comparison between thermal insulation coatings for underwater pipelines. In: Offshore technology conference. Houston

Rønningsen HP (2012) Production of waxy oils on the Norwegian continental shelf: experiences, challenges, and practices. Energy Fuels 26(7):4124e4136. https://doi.org/10.1021/ef300288g

Saffman P (1965) The lift on a small sphere in a slow shear flow. J Fluid Mech 22(2):385e400. https://doi.org/10.1017/S0022112065000824

Singh P, Venkatesan R, Fogler HS (2000a) Formation and aging of incipient thin film wax–oil gels. AIChE J 46:1059–1074

Singh P, Venkatesan R, Fogler HS, Nagarajan N (2000b) Formation and aging of incipient thin film wax-oil gels. AlChE J 46 (5):1059e1074. https://doi.org/10.1002/aic.690460517

Singh P, Venkatesan R, Fogler HS, Nagarajan N (2001) Morphological evolution of thick wax deposits during aging. AIChE J 47(1):6–18

Singhal HK, Sahai GC, Pundeer GS, Chandra K (1991) Designing and selecting wax crystal modifier for optimum field performance based on crude oil composition. In: Annual technical conference and exhibition of the society of petroleum engineers. Society of Petroleum Engineers, Dallas

Suppiah S, Ahmad A, Alderson CJ, Akbarzadeh K, Gao J, Shorthouse JM, Khan IA, Forde C, Jamaluddin AK (2010) Waxy crude production management in a deepwater subsea environment. In: Paper presented at the SPE annual technical conference and exhibition. https://doi.org/10.2118/132615-PA

Taheri-Shakib J, Zojaji I, Saadati N, Kazemzadeh E, Esfandiarian A, RajabiKochi M (2020) Investigating molecular interaction between wax and asphaltene: accounting for wax appearance temperature and crystallization. J Petrol Sci Eng 191:107278. https://doi.org/10.1016/j.petrol.2020.107278

Theyab MA, Diaz P (2017) An experimental and simulation study of wax deposition in hydrocarbon pipeline. Glob J Eng Sci Res 4(7):27–40

Tinsley JF, Prud'homme RK, Guo X, Adamson DH, Shao S, Amin D et al (2007) Effects of polymers on the structure and deposition behavior of waxy oils. In: SPE international symposium on oilfield chemistry. Society of Petroleum Engineers, Houston

Wang KS, Wu CH, Creek JL, Shuler PJ, Tang Y (2003) Evaluation of effects of selected wax inhibitors on paraffin deposition. Petrol Sci Technol 21:369–379
Wang P, Wang W, Gong J, Zhou Y, Yang W (2013) Effects of the dispersed phase on oil/water wax deposition. J Energy Resour Technol 135(4). https://doi.org/10.1115/1.4023932
Xie Y, Chen D, Mai F (2018) Economic pigging cycles for low-throughput pipelines. Adv Mech Eng 10(11):1687814018811198. https://doi.org/10.1177/1687814018811198
Xiong QY, Kiyingi W, Pan JJ, Xiong R, Deng W, Zhang SL, Guo JX, Yang Y (2020a) Analysis of Xinjiang asphaltenes using high precision spectroscopy. RSC Adv 10(65):39425e39433. https://doi.org/10.1039/D0RA07278H
Xiong RY, Guo JX, Kiyingi W, Feng HS, Sun T, Yang X, Li Q (2020b) Method for judging the stability of asphaltenes in crude oil. ACS Omega 5(34):21420e21427. https://doi.org/10.1021/acsomega.0c01779
Zhou G, Xie ZW, Xu XH, Li Q (2020) A new model of overall heat transfer coefficient of hot wax oil pipeline based on dimensionless experimental analysis. Case Stud Therm Eng 20:100647. https://doi.org/10.1016/j.csite.2020.100647
Zhu T, Walker JA, Liang J (2008) Evaluation of wax deposition and its control during production of Alaska North Slope Oils. Final Report, University of Alaska Fairbanks, ConocoPhillips Alaska, Inc, University of Kansas, United States Department of Energy National Energy Technology Laboratory, Oil & Natural Gas Technology, DOE Award No: DE-FC26–01NT41248, p 296

Chapter 2
Challenges During Production, Storage, and Transportation of Waxy and Heavy Crude Oils Due to Organic Deposition

Biswadeep Pal and Tarun Kumar Naiya

Abstract Waxy and heavy crude oils present unique challenges during their production, storage, and transportation due to the phenomenon of organic deposition. This article critically reviews the issues associated with organic deposition in waxy and heavy crude oils and explores the implications for the industry. It discusses the causes and mechanisms of organic deposition, examines its impact on production efficiency, storage capacity, and transportation infrastructure, and evaluates current mitigation strategies. The review highlights the need for proactive measures and innovative technologies to address organic deposition challenges and ensure the smooth and cost-effective handling of waxy and heavy crude oils.

Keywords Waxy crude oil · Heavy crude oil · Organic deposition · Production challenges · Storage limitations · Transportation infrastructure · Mitigation strategies · Technological innovations

2.1 Introduction

The escalating global demand for energy is fuelling extensive exploration and extraction activities within the crude oil industry. However, not all crude oils are the same, and certain types, specifically waxy and heavy crude oils, introduce distinct challenges due to contain high amount of long chain organic hydrocarbon compounds. A repeated and challenging issue associated with these particular crude oils is the deposition of organic substances during their production, storage, and transportation processes.

B. Pal (✉)
Department of Chemical Engineering, Parul University, Vadodara, Gujarat, India
e-mail: pal.biswadeep90@gmail.com

T. K. Naiya
Department of Petroleum Engineering, IIT (ISM), Dhanbad, Jharkhand, India

S. Banerjee et al. (eds.), *Environmental Friendly Green Technologies for Improvement of Heavy Crude Oil Flow Assurance*, Environmental Science and Engineering,
https://doi.org/10.1007/978-3-031-86701-9_2

Crude oil is a complex blend of hydrocarbon compound, with its composition varying significantly depending on its origin. Waxy and heavy crude oils, in particular, contain high levels of organic compounds like paraffin wax, asphaltenes, and resins (Sofer 1984). Under certain conditions, such as fluctuations in temperature, pressure, or chemical composition, these compounds tend to solidify and create deposits on various surfaces, including pipelines, storage tanks, and transportation vessels (Douglas et al. 2022).

The consequences of organic deposition are devasted. These encompass diminished operational efficiency, escalated maintenance expenses, and protect safety risks. Insufficient management of organic deposition can result in disruptions to production schedules, costly shutdowns for equipment cleaning and requiring constant maintenance. In extreme cases, catastrophic incidents that imperil both personnel and the environment (Sotoodeh 2021).

In this chapter, our main focus on the challenges associated with organic deposition in the context of waxy and heavy crude oils. This chapter drives into the distinctive characteristics of these organic compounds, the complications they introduce at various stages of the crude oil lifecycle, and the methodologies used to mitigate these issues. As the industry try to meet new demands and grapple with environmental considerations, comprehending and addressing the intricacies of organic deposition remains pivotal for ensuring the seamless flow of these invaluable energy resources.

2.2 Understanding Organic Deposition

2.2.1 Paraffin Wax Deposition

In the early 1940s, the historical neglect of the paraffin problem in the literature despite its longstanding presence in the oil production industry shows the importance for smooth operations. It mentions the scarcity of comprehensive references on the subject and notes that the most recent thorough examination of the issue is two decades old (Shock et al. 1955). In the present scenario, the issue of paraffin wax deposition is a well-documented and significant challenge within the oil and gas sector, especially in the context of waxy crude oils. The composition of paraffin is varied depends upon its origin (Webber 1999). It is always advisable to understand the importance of paraffin deposition and it is of utmost importance to grasp the mechanisms and ramifications of paraffin wax deposition to implement effective strategies for mitigation. The effectiveness of chemical additives in preventing paraffin deposition varied depending on the type of crude oil being treated. These additives influenced crystal modification, as evidenced by cold flow test results, and correlated with changes in deposition characteristics. While pour point depressants likely achieved inhibition through crystal modification and/or dispersion, asphaltic additives appeared to prevent wax adhesion to pipe walls through adsorption, with

Table 2.1 Advantages for temperature driven transportation

Parameter	Advantage
Separation of impurities	Separating sulphur, water improves the corrosion rate
Wax removal	Allows easy transportation in colder environment
Precipitation-based separation	Enables required specific oil types
Improved product quality	Removal of undesirable components like metals, wax, tar etc.
	Enhancing refined products quality

different fractions of the crude oil's paraffin content affected to varying degrees, as reflected in molecular weight distribution changes (Mendell and Jessen 1970).

2.2.2 Factors Governing Paraffin Wax Deposition

2.2.2.1 Temperature-Driven Precipitation

Paraffin wax consists of a complex mix of long-chain hydrocarbons. At higher temperatures, either within the reservoir or during the transportation phase, paraffin remains in a dissolved state within the crude oil (Leontaritis 1996). However, as temperatures drop, particularly during transportation through pipelines or while being stored in tanks, the wax molecules lose their solubility and transform into solid crystals. This transformation leads to the formation of stubborn wax deposits on various surfaces, including equipment components, pipeline interiors, and even within the crude oil itself (Hassan 2020). The temperature gradient plays a crucial role in wax deposition. Changes in environmental conditions can lead to a decrease in the pipeline surface temperature, affecting the heat transfer between crude oil and the pipeline surface. When the temperature drops below the Wax Appearance Temperature (WAT), wax begins to accumulate and precipitate at the pipeline's bottom. The temperature difference also influences mass transfer rates, with higher deposition rates occurring as the temperature gradient increases. Once the temperature drops below the Pour Point (PP), the flow of Wax-Crude Oil (WCO) ceases. Initially, deposition occurs rapidly because the crude oil comes into direct contact with the cold pipeline surface, resulting in a high temperature difference. However, over time, a wax deposition layer acts as a thermal insulator, reducing the temperature difference. This high temperature difference causes a greater number of paraffins to precipitate, providing sites for larger wax crystal formation (Table 2.1).

2.2.2.2 Pressure and Flow Impact

Alterations in pressure and variations in flow rates can exacerbate the deposition of paraffin wax. Reduced pressure conditions can prompt the sudden precipitation of wax, while changes in flow dynamics can create zones with high shear stress, facilitating the adherence of wax to pipe walls. Pressure decreases as crude oil is transported from the reservoir to surface facilities. After separating the lighter fractions, the solubility of wax in crude oil decreases. This reduction in the WAT results in more light particles being dissolved in the crude oil. These dissolved light particles create pressure differences that directly affect the rate of wax deposition (Joonaki et al. 2020).

2.2.2.3 Experimental Time

The duration of experiments is a critical parameter in wax deposition studies. The early stages of laboratory experiments have a significant impact on wax deposition. In batch processes, crude oil samples do not continuously receive fresh paraffin, so most of the dissolved paraffin is deposited on the wall during the initial hours (Kasumu and Mehrotra 2015). Consequently, the deposition rate is higher in the early hours compared to later stages. In field conditions, wax deposition increases with time as crude oil is continuously supplied during production and gets deposited on the cold surface.

2.2.2.4 Crude Oil Composition

Crude oil consists of both light and heavy hydrocarbon fractions. Light hydrocarbon fractions tend to form solid hydrates at low temperatures and high pressures, while heavy crude oil molecules deposit on the inner pipeline walls at lower temperatures. Consequently, the composition of crude oil has a significant impact on wax deposition (Singhal et al. 1991). However, presence of high amount of wax content results less time to aggregate and form crystal network. Sometimes long chain hydrocarbon made hard paraffin layer which are difficult to remove as a result, appreciable reduction of flowrate was also observed. It is also reported that high amount of water significantly influences the wax deposition by transferring the wax particles from other places due high turbulence. Along with high specific heat capacity, formation of water film on the pipeline surface and remaining gap is filled by water to prevent paraffin deposition.

2.2.2.5 Pipeline Surface Properties

The properties of pipeline surfaces are also influential. Whether the surface is made of carbon steel or stainless steel can affect adhesion rates, as coated surfaces typically reduce friction and discourage deposition. The effective contact area and high friction factor influence adhesion bonds, promoting wax deposition. Consequently, deposition reduces the inner diameter of the pipeline, leading to increased pressure drop, which can eventually block production altogether (Bai et al. 2019).

2.2.3 Mechanism Behind Paraffin Deposition

Shear dispersion due to occurs due to the mechanical shearing of wax molecules caused by the hydrodynamic drag generated by fluid flow. The degree of shear depends primarily on factors such as flow rate and fluid viscosity. Higher viscosity and lower flow rates typically result in increased wax deposition rates (Zolfaghari et al. 2016). However, in situations of highly turbulent flow, higher flow rates can actually reduce deposition rates as the flow's mechanical action shears off wax deposits from surfaces. As the deposit thickness increases, the shear rate rises due to a decrease in flow area and an increase in flow velocity. This higher shear rate exerts increased shear stress on wax molecules and crystals, ultimately decreasing the overall wax deposition rate.

On the other hand, molecular diffusion describes how radial temperature gradients within a pipeline create a concentration gradient of dissolved paraffin components in the liquid phase. This gradient causes paraffin to diffuse toward the metallic surface, leading to deposition. The recognized mechanisms contributing to wax thickness on a metal surface include molecular diffusion of dissolved wax, the transport of precipitated wax particles, and the migration of older wax crystals towards fresh ones on the surface (Ekweribe and Civan 2011).

To enable paraffin deposition from a solution onto a surface, specific temperature conditions must be met. Notably, the temperature of the wax solution must be higher than the temperature of the deposition surface, which must be maintained below the wax solution's cloud point (Gomez et al. 2013).

Freshly precipitated wax crystals typically form or adhere directly to the deposition surface, while older wax crystals, no longer in solution, may migrate towards the fresh ones on the surface and adhere to them. The initial deposition of fresh crystals is a relatively quick process, whereas the latter process is considerably slower. This explains why deposition does not occur when both the wax solution and the deposition surface are below the wax solution's cloud point temperature but the surface is warmer than the solution.

A concentration gradient arises due to the temperature difference between the wax solution and the deposition surface. Although adsorption forces at the metal surface play a role, they are primarily significant only in the initial stages of deposition. Beyond these initial stages, the adhesive forces between wax and metal operate over only a few molecular diameters.

When suspended within oil, small wax crystals experience continuous collisions with thermally agitated oil molecules, leading to random Brownian motion of the suspended particles. In the presence of a concentration gradient of these particles, Brownian motion results in a net transport process similar to diffusion in nature and mathematical description.

The other three mechanisms—Brownian diffusion, gravity settling, and shear dispersion—are considered secondary mechanisms that can contribute to wax deposition. Research suggests that Brownian diffusion of wax particles is unlikely to be a significant factor because the wall temperature is lower than the bulk oil temperature, causing more precipitated wax particles to accumulate at the wall rather than in the bulk oil, thus transporting them away from the wall rather than towards it for deposition. Gravity settling is also believed to be relatively insignificant in wax deposition, as studies have not generally found wax deposits to be thicker at the bottom compared to the top of pipe walls in single-phase oil flow conditions.

Therefore, shear dispersion, molecular diffusion, and other mechanisms collectively influence the complex phenomenon of wax deposition, but the relative significance of each mechanism can vary depending on the specific conditions and characteristics of the system.

2.2.4 Consequences Arising from Paraffin Wax Deposition

The accumulation of paraffin wax within pipelines and equipment hampers the smooth flow of crude oil. This not only diminishes the efficiency of oil transportation but also results in increased energy consumption and the potential for pipeline blockages. The buildup of wax deposits necessitates frequent and costly maintenance procedures. To mitigate these operational challenges upstream industries are employed mechanical or pigging methods for removing organic deposits. Sometimes where the pigging methods is not operatable due to long pipeline length or deep wellbore production situation chemical methods are accepted. In the recent time, some microbes like These may encompass mechanical cleaning processes, the use of chemical treatments, and the deployment of pigging devices aimed at removing the deposits and restoring operational efficiency.

Unaddressed wax deposition can lead to unforeseen operational shutdowns and delays in production. As paraffin wax accumulates on the inner walls of pipelines, it can constrict the flow area, leading to reduced flow rates and increased pressure drop. This reduced flow capacity can result in decreased production and transportation efficiency. Managing paraffin deposition often requires additional operational and maintenance activities. These can include heating the pipeline to prevent wax solidification, pigging operations (mechanical cleaning), and the use of chemical inhibitors. All these activities contribute to increased operational costs.

When pipelines become significantly blocked by wax deposits, the flow of oil or gas may need to be halted to perform extensive cleaning operations, causing production. Preventing paraffin deposition often involves the use of heating systems to maintain the fluid temperature above the wax appearance temperature. This requires additional energy consumption, resulting in higher operating costs and environmental impacts. Wax deposits can create an insulating layer on the pipeline surface, trapping water and other corrosive substances which can accelerate corrosion of the pipeline material beneath the wax layer. Additionally, the mechanical action of wax crystals flowing with the fluid can contribute to erosion of pipeline components.

Routine maintenance and cleaning operations are often required to remove wax deposits from pipelines. These downtime periods can disrupt production schedules and reduce overall system reliability. The commercial polymeric compounds which are used to inhibit paraffin deposition or clean pipelines have environmental impacts. So that, researchers are more focused to ensure bioderived additives which is safe to use and environment friendly.

Paraffin deposits affects the transportation. As a result, viscosity of crude oil changes which also creates problem in downstream industries. Paraffin deposition also occur in reservoirs and wellbores, potentially reducing production rates and causing wellbore obstructions.

2.3 Asphaltene and Resin Deposition

Deposition of asphaltene and resin presents a major challenge in the oil and gas sector, particularly when dealing with heavy and waxy crude oils. Crude oil contains various complex organic compounds, including asphaltenes, resins, and napthalenes, which can cause significant problems during production, storage, and transportation due to their tendency to self-aggregate and form deposits. These deposits can result from changes in temperature, pressure, or chemical composition (Overton et al. 2022).

Asphaltenes are high molecular weight complex organic compounds naturally found in crude oil, characterized by their dark, heavy, and insoluble nature in lighter hydrocarbons. They consist of carbon, hydrogen, oxygen, sulfur, and nitrogen atoms, with intricate aromatic and aliphatic structures that vary depending on the crude oil source.

The deposition of asphaltenes can have severe consequences, including flow obstruction in pipelines, reduced flow rates, increased pressure drop, equipment damage, corrosion, higher maintenance costs, and downtime. Asphaltenes can also affect the quality of refined products, impacting refining processes.

Resins, similar to asphaltenes, are complex organic compounds in crude oil, lighter than asphaltenes but heavier than most other hydrocarbons. Resins act as the "glue" that binds asphaltenes and other solid materials in crude oil. They interact with asphaltenes and other impurities, facilitating their agglomeration and deposition. This interaction can lead to diverse deposits that are challenging to manage and remove (Zaumanis et al. 2014).

Resin deposition can have detrimental consequences, including increased agglomeration of asphaltenes and other impurities, exacerbating deposition issues, and affecting refining processes, ultimately impacting the quality and yield of refined products.

In conclusion, addressing asphaltene and resin deposition challenges in the oil and gas industry requires strategies like temperature and pressure control, chemical treatments, and specialized equipment. Effectively managing these issues is crucial for maintaining operational efficiency, reducing maintenance costs, and ensuring the reliable flow of crude oil from production sites to consumers.

2.3.1 Factors Depends Asphaltene and Resin Deposition

The deposition of asphaltene and resin in crude oil and petroleum products can be influenced by mainly crude oil conditions, pipeline surface, temperature and various other factors. The chemical composition of the crude oil plays a significant role. Crude oils with higher concentrations of asphaltenes and resins are more prone to deposition. Additionally, the type and size of asphaltene molecules can vary between crude oils, affecting their propensity to precipitate. Lower temperatures can lead to the crystallization and precipitation of asphaltenes and resins. Maintaining the oil temperature above the WAT and asphaltene precipitation temperature is crucial to prevent deposition. Changes in pressure can influence the solubility of asphaltenes and resins. Lowering pressure, such as during production or transportation, can lead to the destabilization and precipitation of these components (Rocha et al. 2006). Sometimes high shear rates can disperse or break down deposits, while low shear rates can promote deposition. The presence of polar or non-polar solvents in the crude oil can affect the solubility of asphaltenes and resins. Changes in solvent quality, often due to mixing with other oils or diluents, can trigger deposition. Rapid pressure drops, such as those occurring at choke points in pipelines or during sudden valve closures, can cause asphaltenes and resins to precipitate rapidly (Bukkaraju et al. 2016). These pressure drops create conditions where the oil may become supersaturated with solid components. Deposition occurs especially in heavy and high-pressure reservoirs due to the changes in reservoir temperature and pressure, as well as interactions with formation minerals, can trigger deposition. Moreover, water can interact with asphaltenes and resins, promoting aggregation and precipitation. These interactions may lead to the formation of stable solid structures.

However, Certain materials may promote adhesion and deposition of asphaltenes and resins, especially if they are chemically reactive with these components. Flow rate within pipelines and equipment can affect the degree of turbulence and shear forces experienced by the crude oil. High flow rates may help prevent deposition by promoting dispersion (Table 2.2).

2.3.2 Mechanism Behind Asphaltene and Resin Deposition

Comprehending the mechanisms governing the deposition of heavy organic compounds within the realm of oil production, transportation, and processing is of paramount importance for the advancement of more economical, environmentally friendly, and effective technologies. The nature and scale of heavy organic deposition from petroleum fluids can fluctuate based on the specific hydrocarbons found in the oil and their relative proportions. Four well-established mechanisms can shed light on these depositions, and one or more of these mechanisms might be accountable for organic deposition during different phases of oil operations.

Table 2.2 Factors affecting due to deposition of asphaltene and resin

Consequence	Description
Flow restriction	Gradual buildup of asphaltenes and resins that narrows the pipeline, reducing flow rates and efficiency
Pressure drops	Accumulation of deposits leads to increased pressure difference of the flow system
Plugging of equipment	Deposition in valves, pumps, and other equipment results in blockages
Reduced production efficiency	Flow restrictions and pressure drops hinder oil production rates which affecting overall productivity
Corrosion	Deposition on localized areas prone to corrosion
Increased operational costs	Frequent cleaning, maintenance, and the use of chemicals to address deposition escalate operational expenses
Production downtime	Shutdowns for cleaning and maintenance disrupt production schedules, leading to downtime

2.3.2.1 Poly-dispersivity Effect

The dispersion of heavy organic compounds within oil is influenced by the chemical composition of petroleum. Factors like the ratio of polar to non-polar molecules and the presence of both light and heavy particles in petroleum are crucial for maintaining the stability of the mixed polydisperse oil. Heavy organic deposition can occur when the balanced composition of the polydisperse oil is disrupted. Changes in temperature, pressure, or composition, such as introducing a miscible solvent into the oil, can disturb the polydisperse oil system. This disruption can result in the separation of heavy and/or polar fractions from the oil mixture, leading to the formation of either another liquid phase or solid precipitates. Portions of these separated fractions, which may contain elements like sulfur, nitrogen, oxygen, and hydrogen bonds, can agglomerate and form irreversible heavy organic deposits that may not readily dissolve in other solvents (Kumar et al. 2020).

2.3.2.2 Colloidal Effect

Some heavy organic components, such as asphaltenes, have the potential to separate from the oil phase and agglomerate into larger particles. These particles remain suspended in the oil due to the presence of substances known as peptizing agents, like resins, that adhere to their surfaces. The stability of these colloidal particles is influenced by several factors, including the concentration of the peptizing agent in the solution, the degree to which the peptizing agent covers the surface sites of heavy organic particles, and the equilibrium conditions between the peptizing agent in the solution and on the surface of these particles. The amount of peptizing agent adsorbed primarily relies on its concentration within the oil (Li et al. 2022).

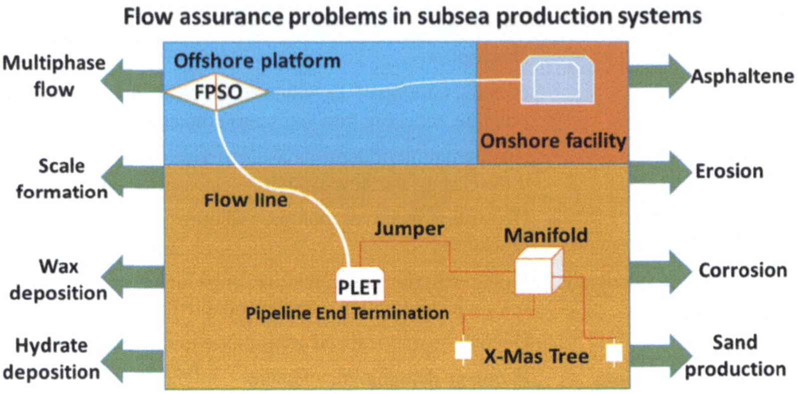

Fig. 2.1 Flow assurance issues faced by oil and gas industries in subsea conditions (Kumar 2023)

2.3.2.3 Aggregation Effect

Adjusting the concentration of a peptizing agent like resin in the oil can impact the amount that adheres to the surface of heavy organic particles. If the concentration decreases to a level where it can't cover the entire surface of these particles, irreversible aggregation may occur. This leads to particles merging, increasing in size, and forming clusters. The properties and structure of these resulting aggregates determine their effect on petroleum fluid behavior. For example, it is believed that asphaltene aggregates adhere to a particular growth pattern (Mansoori 1997).

2.3.2.4 Electrokinetic Effect

When oil flows through conduits like porous media or pipes, the electrokinetic effect becomes important in the deposition of heavy organic components. This effect results from the development of an electrical potential difference along the conduit, caused by the movement of charged particles. This potential difference can modify the behavior of charged and colloidal particles downstream, ultimately leading to their premature deposition and clogging of the conduit (Chilingar and Haroun 2014) (Fig. 2.1).

2.3.3 Consequences Arising from Asphaltene and Resin Deposition

Asphaltene and resin deposition leads to the formation of solid layers on the inner surfaces of pipelines and equipment. These layers reduce the available cross-sectional area for fluid flow. If flow resistance increases, higher pressures are required to maintain the constant flow rates (Ramírez-Jaramillo et al. 2006). The potential for flow instability, such as slug flow or turbulent flow, increases, leading to uneven flow and pressure fluctuations which is shown in Fig. 2.2. Slower flow rates mean it takes longer to transport a given volume of oil from the reservoir to processing facilities or export terminals. Reduced production rates can lead to missed production targets and lower revenue for oil-producing companies (Pirog 2007). Overcoming the elevated resistance caused by deposition requires additional pumping power, resulting in higher operational costs. Operators may need to install larger or more powerful pumps to maintain the desired flow rates. Increased energy consumption for pumping leads to higher electricity costs.

To address severe deposition issues, shutdowns for cleaning is one of most common adopt technologies applied by upstream industries. Shutdown disrupt production schedules and can result in significant downtime. Shutdowns for cleaning are costly in terms of lost production and the expenses associated with maintenance crews and equipment (Chrysalidis and Kyzas 2020).

Corrosion can occur beneath deposits due to the trapped corrosive substances, leading to metal degradation and potential leaks (Zhang et al. 2016). Erosion can result from the abrasive nature of the solid deposits, causing accelerated wear and tear on pipeline surfaces. Asphaltenes and resins can act as emulsifying agents, promoting the formation of stable water-in-oil emulsions (Lee 1999). These emulsions can be challenging to break and can result in higher water content in the crude oil, affecting downstream processing. So that, separating water and oil becomes more difficult, leading to increased operational complexity and reduced product quality. The API gravity of the crude oil may increase, making it lighter and potentially less valuable after deposit into bottom of the pipeline (Briggs et al. 1988). It is also observed

Fig. 2.2 Asphaltene deposit in inner pipeline (Mullins 2008)

that asphaltene binds with resins and wax in cold temperature and get solidified. As a result, viscosity may rise, making the oil harder to transport and process. The presence of solid particles can lead to contamination and affect the quality of refined products.

2.4 Challenges During Production on Organic Deposition

2.4.1 Flow Assurance

Flow assurance is a critical aspect of crude oil production, particularly when dealing with waxy and heavy crude oils. It refers to the management of fluid properties to ensure the continuous and efficient flow of crude oil from the reservoir to processing facilities. Organic deposition, including the formation of wax, asphaltenes, and other solid compounds, presents significant challenges to flow assurance during the production phase (Jamaluddin and Kabir 2012).

The properties of crude oil can change as it travels from the reservoir to surface facilities, where surface conditions differ significantly from reservoir conditions, to the production facilities which is shown. Crude oils are typically produced from deep underground reservoirs, where temperatures reaches high and low pressures. As crude oil is flows to the surface, it encounters lower pressures and temperatures, which can cause various organic compounds within the oil to solidify and deposit on pipeline walls and equipment (Abdulhussein et al. 2023).

2.4.2 Equipment Maintenance

In the complex and demanding realm of oil and gas production, equipment maintenance emerges as a critical aspect of ensuring the uninterrupted flow of crude oil, particularly in the presence of challenges like organic deposition. The efficient operation of production equipment is vital for extracting and transporting crude oil from reservoirs to processing facilities (Aycaguer et al. 2001).

Production facilities in the oil and gas industry consist of an intricate network of machinery, pipelines, pumps, and separators, all working in tandem to extract and transport crude oil. Equipment in these facilities is subjected to harsh environmental conditions, high pressures, varying temperatures, and, in the case of heavy and waxy crude oils, the potential for organic deposition (Elkatory et al. 2022). Therefore, it is crucial to perform routine and thorough equipment maintenance to guarantee safety, operational efficiency, and cost management. Prioritizing the safety of both personnel and the environment is of utmost importance. Properly maintained equipment is less prone to malfunction or experience catastrophic failures, thus minimizing the likelihood of accidents and environmental disasters (Ghorbani et al. 2021).

Well-maintained equipment functions with greater efficiency, reducing operational interruptions and increasing production yields. Regular maintenance serves to detect and resolve potential problems before they escalate into significant issues, averting expensive emergency repairs and equipment replacements. In the oil and gas sector, corrosion is a prevalent concern due to the corrosive properties of crude oil and the presence of corrosive elements in produced water (Parthipan et al. 2017). Corrosion can weaken equipment, leading to leaks and structural failures. Regular inspections and the use of corrosion-resistant materials are essential for mitigating this challenge. The continuous operation of production equipment exposes it to wear and tear. Components like pumps, valves, and pipelines can degrade over time, resulting in reduced performance and reliability. Monitoring and timely replacement of worn parts are crucial for maintaining equipment integrity. Exposure to harsh environmental conditions, such as extreme temperatures, humidity, and corrosive atmospheres, can accelerate equipment deterioration. Protective coatings, insulation, and climate control measures are employed to combat these factors (Aalsalem et al. 2018).

Thus, develop and adhere to a regular maintenance schedule that includes equipment inspections, cleaning, lubrication, and component replacements are essential consideration (Ahuja and Khamba 2008). It is always advisable to utilize advanced sensors and monitoring systems to assess the condition of equipment in real-time, enabling predictive maintenance. Due to high corrosive nature, implementing corrosion management programs, including the use of corrosion-resistant materials, coatings, and cathodic protection. Routinely cleaning and pigging operations have to be initiated to remove organic deposits and debris from pipelines and equipment.

So, equipment maintenance is a pivotal component of crude oil production, especially when dealing with heavy and waxy crude oils that are prone to organic deposition. Addressing corrosion, wear and tear, and other maintenance challenges through proactive strategies is essential for maintaining safe, efficient, and environmentally responsible operations within the oil and gas industry.

2.5 Challenges During Storage on Organic Deposition

2.5.1 Tank Bottom Sediments

When waxy and heavy crude oils are stored for extended periods, sedimentation can occur at the bottom of storage tanks (Carbognani et al. 1999). This sediment, often rich in asphaltenes and resins, can be difficult to remove and can lead to tank corrosion and reduced storage capacity.

Tank bottom sediments, also known as sludge or bottom solids, are accumulations of solid materials that settle at the bottom of storage tanks over time (Jiang et al. 2020; Roffey 1989). These sediments primarily consist of organic compounds such as asphaltenes, resins, and particulate matter, along with water and other impurities present in crude oil (Ali and Alqam 2000). As sediments accumulate, they occupy valuable storage space within the tank, reducing its overall capacity. Sediments can trap water, creating a corrosive environment that accelerates tank bottom corrosion (Sal'nikova et al. 2022). Corrosion can result in tank leaks or structural deterioration (Kokal 2005). Sediments also affect the quality of stored crude oil, making extraction and refining more difficult when necessary. To tackle tank bottom sediments, regular tank cleaning and maintenance are essential to eliminate accumulated sediments, prevent corrosion, and implement effective water removal methods to minimize water content in tanks are required.

2.5.2 Temperature Management

Crude oil storage is a critical aspect of the oil and gas industry, taking place in tanks at various facilities such as production sites, refineries, and distribution terminals. However, challenges related to organic deposition persist during this storage phase, impacting both the integrity of storage tanks and the quality of stored crude oil (Ossai et al. 2020).

Two significant challenges during storage are tank bottom sediments and temperature management. Temperature control is crucial, especially when dealing with waxy and heavy crude oils that contain sensitive organic compounds like paraffin wax and asphaltenes (Sulaimon and Yusoff 2015; Behbahani et al. 2008). Temperature fluctuations can lead to the crystallization of these compounds, resulting in deposits that can impede flow rates during transportation and processing. These temperature variations can also affect the overall quality of stored crude oil, making refining more challenging and potentially causing operational disruptions, increased maintenance, and downtime (Martínez-Palou et al. 2011; Tung et al. 2001).

To address temperature management challenges, specific solutions include implementing heating systems to maintain suitable temperatures and prevent wax crystallization, ensuring proper insulation of storage tanks to minimize heat loss and temperature fluctuations, and implementing continuous temperature monitoring (Jennings and Breitigam 2009).

Thus, crude oil storage poses challenges related to tank bottom sediments and temperature control, particularly when dealing with waxy and heavy crude oils. Effective mitigation strategies, such as regular tank cleaning, water removal, heating systems, insulation, and temperature monitoring, are essential to maintain the quality of stored crude oil and ensure smooth operations within the oil and gas industry (Fawzy et al. 2020).

2.6 Challenges During Transportation on Organic Deposition

2.6.1 *Pipeline Integrity*

Efficient and secure transportation of crude oil is a critical aspect of the oil and gas industry, but it comes with significant challenges, particularly when dealing with heavy and waxy crude oils. Two major issues that arise during transportation are the maintenance of pipeline integrity and the management of viscosity for effective pumping (Priyanka et al. 2021).

Preserving the integrity of pipelines is crucial to ensure the safe and reliable transportation of crude oil. Pipelines act as the lifelines of the oil and gas sector, spanning vast distances and traversing various terrains, including oceans, deserts, and urban areas (Revie 2015). Any compromise in pipeline integrity can lead to catastrophic consequences, including oil spills, environmental damage, and safety hazards.

Pipelines are susceptible to corrosion due to the corrosive nature of crude oil and external factors such as soil composition and water content. Corrosion weakens the walls of pipelines, making them vulnerable to leaks and ruptures (Wasim and Djukic 2022). As previously discussed, organic deposition, including substances like asphaltenes and waxes, can accumulate on the inner walls of pipelines. This buildup reduces the pipeline's internal diameter and increases friction, ultimately resulting in flow restrictions and reduced capacity (Bern et al. 1980). Pipelines face external stressors, such as ground movements, seismic activity, and third-party damage (e.g., construction accidents) (Nyman et al. 2008). These external factors can compromise the integrity of pipelines.

2.6.2 *Viscosity and Pumping*

Heavy crude oils typically possess higher viscosity compared to lighter crude oils. Elevated viscosity can necessitate greater pumping power to propel the oil through pipelines, resulting in increased energy consumption and potential equipment wear (Quarini and Shire 2007; Dunn et al. 1995). Viscosity is highly sensitive to temperature which is shown in Fig. 2.3. Heavy and waxy crude oils may solidify or become more viscous at lower temperatures, potentially leading to blockages within pipelines. Some organic materials, like heavy crude oil or bitumen, have high viscosities at ambient temperatures. High viscosity makes it difficult to pump and transport these materials through pipelines or other conduits (Faris et al. 2015). Maintaining a consistent temperature throughout the transportation process is critical to preventing the material from solidifying or becoming too viscous.

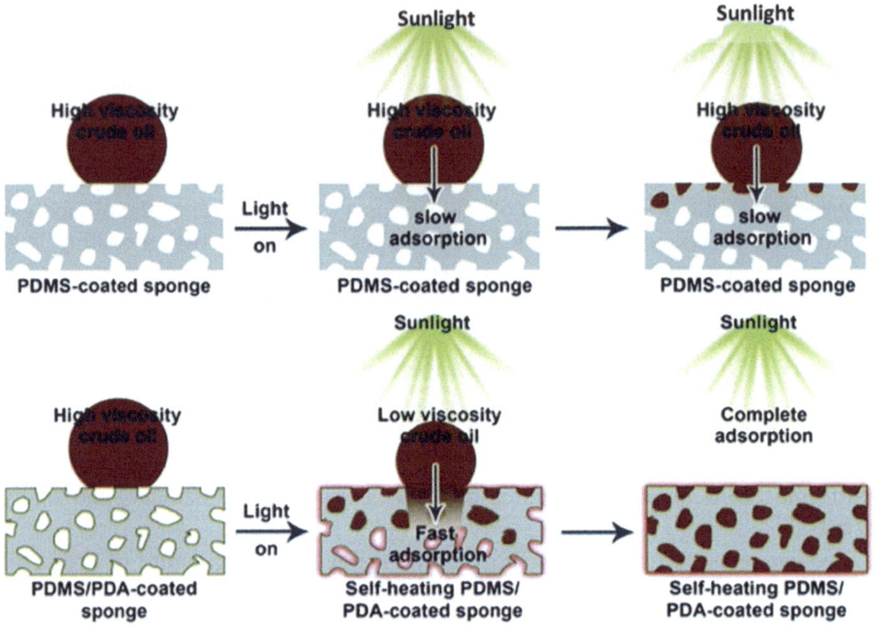

Fig. 2.3 Wettability behaviour with the effect of temperature (Zhang et al. 2018)

2.7 Conclusion

Organic deposition remains a significant challenge during the production, storage, and transportation of waxy and heavy crude oils. Understanding the properties of these organic compounds and implementing effective mitigation strategies is crucial for ensuring the efficient and safe operation of oil and gas facilities. As the industry continues to evolve, ongoing research and development efforts are essential to develop innovative solutions to address organic deposition challenges effectively.

References

Aalsalem MY, Khan WZ, Gharibi W, Khan MK, Arshad Q (2018) Wireless sensor networks in oil and gas industry: recent advances, taxonomy, requirements, and open challenges. J Netw Comput Appl 113:87–97. https://doi.org/10.1016/j.jnca.2018.04.004

Abdulhussein ZA, Al-Sharify ZT, Alzuraiji M (2023) Flow of crude oil in pipes and its environmental impact: a review. Nucl Atmos Aerosols. https://doi.org/10.1063/5.0150151

Ahuja IS, Khamba JS (2008) Total productive maintenance: literature review and directions. Int J Qual Reliab Manag 25(7):709–756. https://doi.org/10.1108/02656710810890890

Ali MF, Alqam MH (2000) The role of asphaltenes, resins and other solids in the stabilization of water in oil emulsions and its effects on oil production in Saudi oil fields. Fuel 79(11):1309–1316. https://doi.org/10.1016/s0016-2361(99)00268-9

Aycaguer A, Lev-On M, Winer AM (2001) Reducing carbon dioxide emissions with enhanced oil recovery projects: a life cycle assessment approach. Energy Fuels 15(2):303–308. https://doi.org/10.1021/ef000258a

Bai J, Xu J, Wu J (2019) Multifunctional anti-wax coatings for paraffin control in oil pipelines. Pet Sci 16(3):619–631. https://doi.org/10.1007/s12182-019-0309-7

Behbahani TJ, Rahmatollah G, Hassan A, Alireza D (2008) Effect of wax inhibitors on pour point and rheological properties of Iranian waxy crude oil. Fuel Process Technol 89(10):973–977. https://doi.org/10.1016/j.fuproc.2008.03.013

Bern P, Withers V, Cairns RJ (1980) Wax deposition in crude oil pipelines. All Days. https://doi.org/10.2118/206-1980-ms

Briggs P, Baron P, Fulleylove R, Wright NJ (1988) Development of heavy-oil reservoirs. J Petrol Technol 40(02):206–214. https://doi.org/10.2118/15748-pa

Bukkaraju SK, Osorio NF, Annadorai KM, Garduno J, Golden NK (2016) Practical guidelines for the diagnosis and remediation of pipeline blockages. In: Offshore technology conference, SPE. https://doi.org/10.4043/27001-ms

Carbognani L, Orea M, Fonseca MG (1999) Complex nature of separated solid phases from crude oils. Energy Fuels 13(2):351–358. https://doi.org/10.1021/ef9801975

Chilingar GV, Haroun M (2014) Electrokinetics for petroleum and environmental engineers. Wiley eBooks. https://doi.org/10.1002/9781118842805

Chrysalidis A, Kyzas GZ (2020) Applied cleaning methods of oil residues from industrial tanks. Processes 8(5):569. https://doi.org/10.3390/pr8050569

Douglas LD, Rivera-Gonzalez N, Cool N, Bajpayee A, Udayakantha M, Liu GW, Anita, Banerjee S (2022) A materials science perspective of midstream challenges in the utilization of heavy crude oil. ACS Omega 7(2):1547–1574. https://doi.org/10.1021/acsomega.1c06399

Dunn L, Matthews C, Zahacy TA (1995) Progressing cavity pumping system applications in heavy oil production. All Days. https://doi.org/10.2118/30271-ms

Ekweribe C, Civan F (2011) Transient wax gel formation model for shut-in subsea pipelines. J Energy Resour Technol Trans ASME 133(3). https://doi.org/10.1115/1.4004965

Elkatory MR, Soliman EA, Nemr AE, Hassaan MA, Ragab S, El-Nemr MA, Pantaleo A (2022) Mitigation and remediation technologies of waxy crude oils' deposition within transportation pipelines: a review. Polymers 14(16):3231. https://doi.org/10.3390/polym14163231

Faris HA, Sami NA, Abdulrazak AA, Sangwai JS (2015) The performance of toluene and naphtha as viscosity and drag reducing solvents for the pipeline transportation of heavy crude oil. Pet Sci Technol 33(8):952–960. https://doi.org/10.1080/10916466.2015.1030079

Fawzy S, Osman AI, Doran J, Rooney D (2020) Strategies for mitigation of climate change: a review. Environ Chem Lett 18(6):2069–2094. https://doi.org/10.1007/s10311-020-01059-w

Ghorbani H, Researchers Y, Abdali MR, Mohamadian N, Wood DA, Limited LDE (2021) Petroleum well blowouts as a threat to drilling operation and wellbore sustainability: causes, prevention, safety and emergency response. J Constr Mater. https://doi.org/10.36756/jcm.si1.1r

Hassan KYA (2020) Descaling petroleum production tubing using multiple high-pressure nozzles. https://salford-repository.worktribe.com/output/1360054

Jamaluddin A, Kabir C (2012) Flow assurance: managing flow dynamics and production chemistry. J Petrol Sci Eng 100:106–116. https://doi.org/10.1016/j.petrol.2012.11.017

Jennings DW, Breitigam J (2009) Paraffin inhibitor formulations for different application environments: from heated injection in the desert to extreme cold arctic temperatures. In: Presented at the 10th international conference on petroleum phase behavior and fouling. Energy Fuels 24(4):2337–2349. https://doi.org/10.1021/ef900972u

Jiang H, Wang Y, Nie C, Yan F, Ouyang X, Gong J (2020) Oil sludge deposition in storage tanks: a case study for Russian crude oil in Mo-he station. Appl Sci 11(1):321. https://doi.org/10.3390/app11010321

Joonaki E, Hassanpouryouzband A, Burgass RW, Hase A, Tohidi B (2020) Effects of waxes and the related chemicals on asphaltene aggregation and deposition phenomena: experimental and modeling studies. ACS Omega 5(13):7124–7134. https://doi.org/10.1021/acsomega.9b03460

Kasumu AS, Mehrotra AK (2015) Solids deposition from wax–solvent–water “waxy” mixtures using a cold finger apparatus. Energy Fuels 29(2):501–511. https://doi.org/10.1021/ef501835b
Kokal S (2005) Crude-oil emulsions: a state-of-the-art review. SPE Prod Facil 20(01):5–13. https://doi.org/10.2118/77497-pa
Kumar A (2023) Perspectives of flow assurance problems in oil and gas production: a mini-review. Energy Fuels 37(12):8142–8159. https://doi.org/10.1021/acs.energyfuels.3c00843
Kumar A, Kuang Y, Liang Z, Sun X (2020) Microwave chemistry, recent advancements, and eco-friendly microwave-assisted synthesis of nanoarchitectures and their applications: a review. Mater Today Nano 11:100076. https://doi.org/10.1016/j.mtnano.2020.100076
Lee RF (1999) Agents which promote and stabilize water-in-oil emulsions. Spill Sci Technol Bull 5(2):117–126. https://doi.org/10.1016/s1353-2561(98)00028-0
Leontaritis KJ (1996) The asphaltene and wax deposition envelopes. Fuel Sci Technol Int 14(1–2):13–39. https://doi.org/10.1080/08843759608947560
Li C, Ma F, Fu Z, Dai J, Wen Y, Shi K (2022) Using cereclor plasticizer to modify the virgin asphalt binder: a case of rheological properties improvement. Constr Build Mater 318:126039. https://doi.org/10.1016/j.conbuildmat.2021.126039
Mansoori GA (1997) Modeling of asphaltene and other heavy organic depositions. J Petrol Sci Eng 17(1–2):101–111. https://doi.org/10.1016/s0920-4105(96)00059-9
Martínez-Palou R, De Lourdes Mosqueira M, Zapata-Rendón B, Juárez EM, Bernal-Huicochea C, De La Cruz Clavel-Lopez J, Aburto J (2011) Transportation of heavy and extra-heavy crude oil by pipeline: a review. J Petrol Sci Eng 75(3–4):274–282. https://doi.org/10.1016/j.petrol.2010.11.020
Mendell JL, Jessen FW (1970) Mechanism of inhibition of paraffin deposition in crude oil systems. In: SPE production techniques symposium, pp SPE-2868. SPE. https://doi.org/10.2118/2868-MS
Nyman DJ, Lee EM, Audibert JM (2008) Mitigating geohazards for international pipeline projects: challenges and lessons learned. In: International pipeline conference, ASME 48593, pp 639–648. https://doi.org/10.1115/ipc2008-64405
Ossai IC, Ahmed A, Hassan A, Hamid FS (2020) Remediation of soil and water contaminated with petroleum hydrocarbon: a review. Environ Technol Innov 17:100526. https://doi.org/10.1016/j.eti.2019.100526
Overton EB, Adhikari PL, Radović JR, Passow U (2022) Fates of petroleum during the deepwater horizon oil spill: a chemistry perspective. Front Mar Sci 9. https://doi.org/10.3389/fmars.2022.928576
Parthipan P, Babu TG, Anandkumar B, Rajasekar A (2017) Biocorrosion and its impact on carbon steel API 5LX by Bacillus subtilis A1 and Bacillus cereus A4 isolated from Indian crude oil reservoir. J Bio Tribo Corros 3(3). https://doi.org/10.1007/s40735-017-0091-2
Pirog RL (2007) Role of national oil companies in the international oil market [August 21, 2007]. Library of congress. Congr Res Serv. https://www.hsdl.org/?abstract&did=717017
Priyanka EB, Thangavel S, Gao X (2021) Review analysis on cloud computing based smart grid technology in the oil pipeline sensor network system. Petrol Res 6(1):77–90. https://doi.org/10.1016/j.ptlrs.2020.10.001
Quarini J, Shire S (2007) A review of fluid-driven pipeline pigs and their applications. Proc Instit Mech Eng Part E J Process Mech Eng 221(1):1–10. https://doi.org/10.1243/0954408jpme108
Ramírez-Jaramillo E, Lira-Galeana C, Manero O (2006) Modeling asphaltene deposition in production pipelines. Energy Fuels 20(3):1184–1196. https://doi.org/10.1021/ef050262s
Rocha LC, Ferreira MS, Da Silva Ramos AC (2006) Inhibition of asphaltene precipitation in Brazilian crude oils using new oil soluble amphiphiles. J Petrol Sci Eng 51(1–2):26–36. https://doi.org/10.1016/j.petrol.2005.11.006
Roffey R (1989) Microbial problems during long-term storage of petroleum products underground in rock caverns. Int Biodeterior 25(1–3):219–236. https://doi.org/10.1016/0265-3036(89)90048-1

Sal'nikova TV, Tumanyan BP, Власова ГВ, Пивоварова НА, Kozyrev ON, Kolosov VM (2022) Deposit formation in equipment for the petroleum and natural gas industry. Chem Technol Fuels Oils 58(3):493–501. https://doi.org/10.1007/s10553-022-01413-4

Shock DA, Sudbury JD, Crockett JJ (1955) Studies of the mechanism of paraffin deposition and its control. J Petrol Technol 7(09):23–28. https://doi.org/10.2118/384-G

Sofer Z (1984) Stable carbon isotope compositions of crude oils: application to source depositional environments and petroleum alteration. AAPG Bull 68(1):31–49. https://doi.org/10.1306/AD460963-16F7-11D7-8645000102C1865D

Sotoodeh K (2021) Prevention of actuator emissions in the oil and gas industry. Gulf Professional Publishing. ISBN = 9780323919173

Sulaimon AA, Yusoff MH (2015) Wax and asphaltene deposition tendency of Malaysian crude oils. Springer eBooks, pp 155–169. https://doi.org/10.1007/978-981-287-368-2_15

Tung NP, Van Vuong N, Long BQK, Vinh NQ, Hung PV, Hue VT, Hoe LD (2001) Studying the mechanism of magnetic field influence on paraffin crude oil viscosity and wax deposition reductions. All Days. https://doi.org/10.2118/68749-ms

Wasim M, Djukic MB (2022) External corrosion of oil and gas pipelines: a review of failure mechanisms and predictive preventions. J Nat Gas Sci Eng 100:104467. https://doi.org/10.1016/j.jngse.2022.104467

Webber RM (1999) Low temperature rheology of lubricating mineral oils: effects of cooling rate and wax crystallization on flow properties of base oils. J Rheol 43(4):911–931. https://doi.org/10.1122/1.551045

Zaumanis M, Mallick RB, Poulikakos LD, Frank R (2014) Influence of six rejuvenators on the performance properties of Reclaimed Asphalt Pavement (RAP) binder and 100% recycled asphalt mixtures. Constr Build Mater 71:538–550. https://doi.org/10.1016/j.conbuildmat.2014.08.073

Zhang D, Qian H, Wang L, Li Y (2016) Comparison of barrier properties for a superhydrophobic epoxy coating under different simulated corrosion environments. Corros Sci 103:230–241. https://doi.org/10.1016/j.corsci.2015.11.023

Zhang C, Wu M, Wu B, Yang J, Xu Z (2018) Solar-driven self-heating sponges for highly efficient crude oil spill remediation. J Mater Chem A Mater Energy Sustain 6(19):8880–8885. https://doi.org/10.1039/c8ta02336k

Zolfaghari R, Fakhru'l-Razi A, Abdullah LC, Elnashaie S, Pendashteh A (2016) Demulsification techniques of water-in-oil and oil-in-water emulsions in petroleum industry. Sep Purif Technol 170:377–407. https://doi.org/10.1016/j.seppur.2016.06.026

Chapter 3
Renewable and Eco-Friendly Biosurfactants for Potential Applications in Enhanced Oil Recovery: Current State and Future Perspectives

Md. Merajul Islam, Neha Saxena, Amit Kumar, and Prathibha Pillai

Abstract Due to rising energy demands and limited fossil fuel supplies, countries throughout the world are looking at boosting the use of alternative energy sources or increasing production from existing reserves. Surfactant injection is a well-known technology for enhancing oil recovery (EOR) because it decreases the interfacial tension (IFT) between oil and water, hence increasing oil output. However, there are also some disadvantages to employing surfactants, such as their availability from non-renewable resources, their high cost, their influence on humans and other creatures due to their toxicological potential, and their negative environmental implications. Biosurfactants reduce the surface tension (ST) of a liquid phase and the interfacial tension (IFT) of two different phases; they are microbiological surface-active agents. They are biotechnological products of high value because of their many uses, low toxicity, ease of preparation, and specific performance in fields as varied as organic chemicals and fertilisers, agrochemicals, metallurgy and mining, cosmetics, foods, medicine, pharmaceuticals, beverages, environmental management, and petroleum and petrochemical applications in emulsifying and demulsifying wetting agents, detergent spreading, and foaming agents. Since bacteria are responsible for biosurfactant synthesis, the wide range of genetic diversity among microbes provides a robust capacity to manufacture novel biosurfactants that can advance EOR technology.

Keywords Biosurfactants · EOR · Eco-friendly · Renewable · IFT · Microbes

Md. M. Islam (✉) · N. Saxena
School of Basic Sciences & Technology, IIMT University, Meerut, Uttar Pradesh, India
e-mail: merajchem29@gmail.com

A. Kumar
Department of Petroleum Engineering and Geoengineering, Rajiv Gandhi Institute of Petroleum Technology, Jais, Uttar Pradesh, India

P. Pillai
Energy Cluster UPES, UPES, Dehradun, Uttarakhand, India

S. Banerjee et al. (eds.), *Environmental Friendly Green Technologies for Improvement of Heavy Crude Oil Flow Assurance*, Environmental Science and Engineering,
https://doi.org/10.1007/978-3-031-86701-9_3

3.1 Introduction

In contemporary society, the existence of human beings is significantly reliant on fossil fuel and its associated byproducts, including kerosene, petrol, petrol and diesel (Geetha et al. 2018a, b; Xi et al. 2021). Crude oil has been a crucial component in the industrial revolution throughout the course of human civilisation, serving as a primary source of global energy supplies (Geetha et al. 2018a, b). The statistical data pertaining to the usage of global oil resources demonstrates a discernible upward trajectory from the previous century up through the year 2019. In 2019, the consumption of oil as a primary energy source amounted to approximately 36,390.5 million barrels (Mbbl). It is anticipated that there will be a significant increase in oil consumption in the future, leading to a corresponding rise in oil prices (Schulz 2016). Figure 3.1 illustrates the temporal evolution of global crude oil consumption from 1980 to the present, while also projecting an upward trajectory in future consumption demand (Soñichsen 2021). The graph exhibited a decline in 2020 as a consequence of the global COVID-19 pandemic and extensive closures; yet, a surge in global energy consumption is anticipated in the forthcoming decades (Soñichsen 2021). The increasing global energy demands, coupled with the finite availability of fossil fuel supplies, necessitate a strong and comprehensive approach to addressing energy supply. The enhancement and advancement of the oil extraction methodology for current reservoirs, together with the identification of alternate resources, represent the most effective approach to mitigate reliance on fossil fuels in the forthcoming years (Ardabili et al. 2019; Nikolova and Gutierrez 2021; Xi et al. 2021).

There are various strategies that are widely recognised and studied for the purpose of improving oil recovery from depleted reserves (Al-Sulaimani et al. 2011). The process of extracting oil from reservoirs consists of primary, secondary, and tertiary recovery stages (Bera and Belhaj 2016; Kokal and Al-Kaabi 2010). During the early period of recovery, oil production is primarily influenced by the reservoir's initial pressure energy, resulting in the extraction of just a fraction, around 10–20%, of the

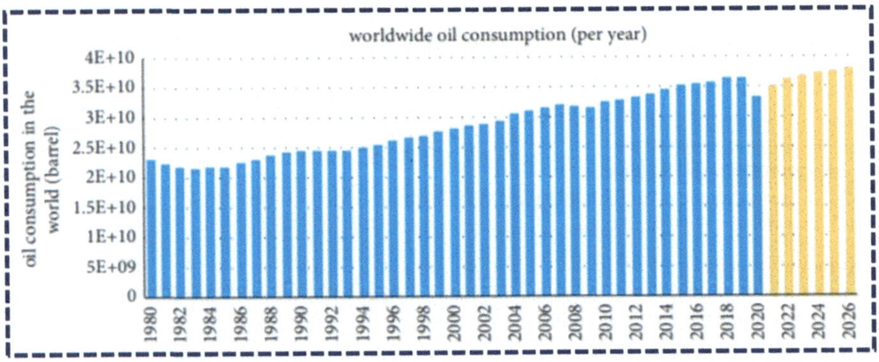

Fig. 3.1 Oil resources used annually around the world (in barrels) (Soñichsen 2021)

current oil reserves within the reservoir (Bera et al 2014). During the last phase of recovery, either gas or water will be introduced into the reservoir to sustain optimal pressures for both the reservoir and oil during the production process. This final phase of oil recovery has been observed to yield an increase of 20–40% (Al-Sulaimani et al. 2011; Xi et al. 2021). In order to ascertain the levels of oil and water saturation relative to each other, it is crucial to carefully select the appropriate injection fluid, consider various chemical factors, and take into account the prevailing reservoir conditions. These conditions encompass factors such as pressure, temperature, brine characterization, pollution levels, rock type and sizes, as well as the distribution of pore structure information (Owens and Archer 1971). If the viscosity of the injected fluid is lower than that of the displacing fluid, it is possible for the injected compound to exhibit a higher flow rate compared to the initial compound's flow through the porous media (Holmberg et al. 2002). The production of oil, both initially and through secondary methods, is influenced by factors such as the characteristics of the oil, geographical challenges, and conditions within the reservoir. These methods have the potential to extract 40–60% of the original oil in place (OOIP). Furthermore, the residual oil saturation (Sor) is a specific objective in enhanced oil recovery (EOR) techniques (Al-Sulaimani et al. 2011; Xi et al. 2021; Park et al. 2015). The optimal period for implementing enhanced oil recovery (EOR) techniques in an oilfield appears to be during the initial phases following primary recovery. Nevertheless, the specific timing is typically contingent upon the economic considerations inherent in the production process (Park et al. 2015). Figure 3.2 provides a comprehensive depiction of the field development programme, illustrating the three distinct manufacturing stages.

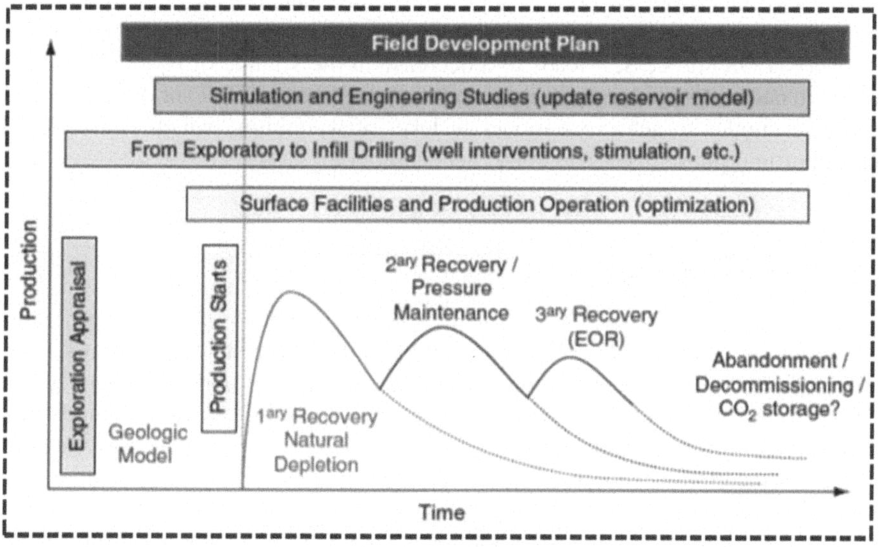

Fig. 3.2 Increasing the amount of oil that is produced over time (Bachari et al. 2019)

The combination of a high interfacial tension (IFT) between oil and water, along with a significant capillary force and electrostatic charge, results in the entrapment of a substantial quantity of unrecovered oil within the pores (Bachari et al. 2019; Xi et al. 2021; Seethepalli et al. 2004). Enhanced oil recovery (EOR) techniques have been implemented during the secondary or tertiary stages to enhance oil recovery yields and reduce residual oil value (Lake 2014). One notable distinction between enhanced oil recovery (EOR) and conventional reservoir recovery methods, such as water and gas injection, lies in the ability of EOR to effectively mobilise the residual oil present in the porous media. This mobilisation leads to more economically favourable outcomes, as it extends the lifespan of reservoirs and prolongs the production of crude oil (Sabhapondit et al. 2003). The inverse Fourier transform of the injecting fluid plays a pivotal role in enhanced oil recovery (EOR). The significant interfacial tension (IFT) observed between injected water and oil hinders the displacement of oil, leading to a substantial accumulation of unrecovered oil within reservoirs. In the interim, the reduction in interfacial tension (IFT) resulting from the implementation of enhanced oil recovery (EOR) fluid serves to diminish the quantity of unrecovered oil and improve the effectiveness of microscopic sweep, as evidenced by prior studies (Holmberg et al. 2002; Ahmadi et al 2014; Xi et al. 2021). One additional factor contributing to the limited extraction of oil and its retention within the pore spaces is the presence of capillary forces. These forces can be effectively mitigated through the implementation of enhanced oil recovery (EOR) techniques (Ahmadi et al 2014).

In order to illustrate the influence of capillary forces, researchers have examined the non-dimensional capillary number, which is defined as the ratio of fluid viscosity (μ) to velocity (V), interfacial tension (IFT) between oil and water (σ), and contact angle (θ) (Chau 2009; Xi et al. 2021). As the aforementioned ratio experiences an increase of four to five orders of magnitude in the context of an Enhanced Oil Recovery (EOR) process, it leads to a decrease in the volume of residual oil. This reduction necessitates either a decrease in Interfacial Tension (IFT) or the adjustment of contact angles to achieve a moderate level of wettability, which is around 90 degrees of interphase (θ) (Chau 2009; Schechter et al. 1991).

3.2 Biosurfactants

Biosurfactants encompass a collection of amphiphilic chemicals that exhibit various biological actions, including antiviral, hemolytic, insecticidal, and antibacterial properties. These compounds have wide-ranging applications across various industries, such as the food industry, cleaning product manufacturing, pharmacology (specifically as medication solvents), cosmetics production, pesticide development, textile manufacturing, organic surface fungicides, medicine, and the oil and gas sector. These applications highlight their significance as vital biotechnology products (Xi et al. 2021). The increasing attention towards biosurfactants can be attributed to their little environmental footprint, low levels of toxicity, and ability to undergo biodegradation (Desai and Banat 1997). The worldwide biosurfactants market achieved a value of 1.5 billion USD by 2019 and is projected to exhibit a compound annual growth rate (CAGR) above 5.5% by 2026. It is noteworthy that Europe accounts for more than half of this market share (Nikolova and Gutierrez 2021). The highest application market is comprised of household detergents, cosmetics, personal care products, and the food industry (Nikolova and Gutierrez 2021; Singh et al. 2019).

The initial synthesis of biosurfactants occurred in 1960 through the utilisation of bacteria in hydrocarbon fermentation, resulting in the production of extracellular molecules (Sari et al. 2019). Biofilms are formed on living surfaces and have the ability to decrease surface tension (ST) and interfacial tension (IFT). The most optimal locations for their synthesis are the extracellular hydrophilic and hydrophobic regions or surfaces of microbial cells (Özdemir et al. 2004). As a result of increased recognition of the detrimental impacts of chemical surfactants on the environment, there has been a change in focus towards surfactants that are more environmentally friendly (Desai and Banat 1997). Several parameters have been identified as affecting the production of biosurfactants, including carbon content, pH, the kind of nitrogen supply, temperature, aeration, and carbon-to-nitrogen ratio. The primary firms involved in the production of biosurfactants include Evonik, Jeneil Biotech, Biotensidon, and Ecover. Additionally, the genera *Pseudomonas, Bacillus, Sphingomonas,* and *Actinobacteria* are recognised as prominent producers of biosurfactant compounds. Desai and Banat (1997) conducted an extensive examination of biosurfactants, with our discussion limited to those biosurfactants that exhibit potential for enhancing oil recovery, namely glycolipids, rhamnolipids, lipopeptides, and sophorolipids.

In a study conducted by Ozdemir et al. (2004), a comparison was made between the lowering of interfacial tension in two distinct solutions of pure rhamnolipids. It was discovered that the molecules of rhamnolipids exhibit robust intermolecular interactions, leading to exceptional foam-formation characteristics, even when subjected to a limited air flow rate. Nevertheless, it should be noted that some yeast-derived biosurfactants, such as sophorolipids and glycolipids, exhibited only moderate levels of foaming capabilities.

The efficacy of sophorolipids (SLs) derived from nonedible Jatropha oil in stain removal was examined by Joshi-Navare et al. (2013). The researchers conducted an optimisation of fermentation settings in order to get the highest possible yield of Jatropha oil-derived saponin-like compounds (SLJO). The results of this optimisation showed a notable reduction in surface tension (ST) when the SLJO was added to distilled water. Furthermore, the SLJO exhibited antibacterial properties and indicated effectiveness in eliminating stains. Additionally, the SLJO displayed remarkable stability as an emulsion when subjected to variations in pH and temperature. The decreased duration of immersion in the washing process, coupled with their antibacterial efficacy, biodegradability, and skin-friendly characteristics, render them very suitable substitutes for synthetic surfactants in household detergents, thereby mitigating their negative impacts.

In their research, Sajna et al. (2013) conducted a study on a biosurfactant derived from a *Pseudozyma* sp. NII08165. This biosurfactant was found to consist of an unidentified glycolipid in conjunction with three isomers of mannosylerythritol lipid

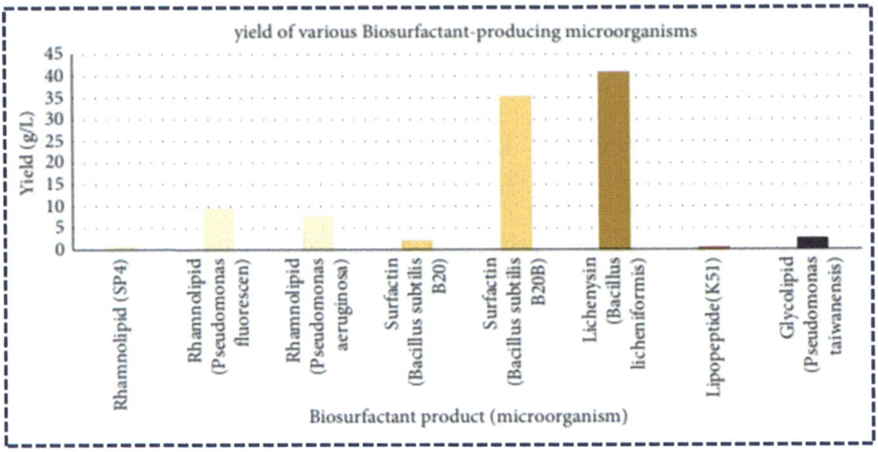

Fig. 3.3 Biosurfactant production by different microorganisms (Sari et al. 2019; Xi et al. 2021)

(MEL). Due to their capacity to maintain stability throughout the alkaline pH range and withstand high temperatures, these substances exhibit suitability as additions in laundry detergents, enabling efficient stain removal. Figure 3.3 presents a comparative analysis of the yield exhibited by different microorganisms engaged in the production of biosurfactants.

3.2.1 Types of Biosurfactants

Biosurfactants can be defined as compounds that are produced by biological sources such as microorganisms, plants, or animals, and possess surface-active properties due to their amphiphilic nature. Microorganisms, including bacteria, yeast, and archaea, are widely recognised as an economically exciting and environmentally friendly source of surface-active compounds. This is primarily due to their extensive genomic diversity, which offers significant potential for various applications. According to Shahaliyan et al. (2015), these compounds have the ability to either become a component of a microorganism's cell wall or be released outside of the cell. Biosurfactants have the ability to dissolve in both polar and non-polar solvents due to their amphiphilic nature (Chen et al. 2015). The effectiveness of a surfactant can be evaluated based on its capacity to reduce the surface tension (ST) and interfacial tension (IFT) between two phases that do not mix well, such as air/liquid and non-polar/polar liquids. The surface tension (ST) is a quantitative measure of the amount of energy needed to expand the surface area of a liquid, taking into account the intermolecular forces involved. The presence of a surfactant reduces the amount of work needed to bring a molecule to the surface, resulting in a decrease in surface tension. According to Mulligan (2005), it is widely acknowledged that an effective biosurfactant has the ability to significantly reduce the surface tension (ST) of water, bringing it down from 72 to less than 35 mN/m. Additionally, it can also decrease the interfacial tension (IFT) between water and n-hexadecane, reducing it from 40 to 1 mN/m. A biosurfactant is considered efficient if it possesses a low critical micelle concentration (CMC). The critical micelle concentration (CMC) is the minimum concentration needed to start forming micelles. It is typically related to the surface tension (ST) and interfacial tension (IFT). In simpler terms, a low CMC indicates that a smaller amount of biosurfactant is needed to decrease the ST or IFT.

Different microbial species produce biosurfactants with widely varying chemical compositions, which can be roughly categorised according to molecular weight or chemical charge. Low-molecular-weight (LMW) surfactants lower the surface tension among two immiscible liquids, while high-molecular-weight (HMW) emulsifiers allow for the creation of oil-in-water or water-in-oil emulsions; HMW emulsifiers are also known as polymeric surfactants (or bioemulsifiers) and are typically made up of exopolysaccharides (EPS) (Nikolova and Gutierrez 2021).

The primary chemical structures of LMW biosurfactants include glycolipids, phospholipids, fatty acids, lipopeptides, and lipoproteins. The formation of biosurfactant can occur in various structures, including single macromolecules, polymers, and particulate structures (Banat 1995; Makkar and Rockne 2003; Satpute et al. 2010). The primary role of LMW is to act as surface-active agents, which involves reducing the surface and/or interfacial tension between two phases that are not capable of mixing, such as liquid and liquid, liquid and solid, or liquid and gas. Glycolipids and lipopeptides are two types of biosurfactants that have been extensively researched. Glycolipids are composed of various sugars connected to β-hydroxy fatty acids, forming a carbohydrate head and a lipid tail. On the other hand, lipopeptides are made up of cycloheptapeptides with amino acids attached to fatty acids of varying chain lengths. These details are summarised in Table 3.1 of the study by Uzoigwe et al. (2015).

High-molecular weight bioemulsifiers are characterised by their complexity, as they are composed of a combination of heteropolysaccharides, lipopolysaccharides, lipoproteins, and proteins. The term "EPS" is another name by which they are commonly referred. EPS molecules, like low-molecular weight biosurfactants, have the ability to effectively emulsify two liquids that do not mix well, such as oil and water. However, they are not as efficient at reducing surface tension compared to low-molecular weight biosurfactants. EPS molecules exhibit a strong affinity for dispersed hydrocarbons in oil polluted environments, effectively inhibiting the coalescence and rupture of oil droplets. The mechanism being referred to is called stabilisation of emulsion. It is believed to occur due to the presence of numerous reactive groups in the structures (Uzoigwe et al. 2015). Table 3.1 presents some of the most extensively researched microbial EPS used for EOR/MEOR.

Table 3.1 Table of biosurfactants organised by their molecular weight, origin, and EOR/MEOR application (Uzoigwe et al. 2015; Nikolova and Gutierrez 2021)

Biosurfactant type	Origin	Chemical structure	Industrial application	References
Low—molecular weight biosurfactants				
Lipopeptides Surfactin, iturin and fengycin	*Bacillus subtilis*	Cyclic lipopeptide consisting of long hydroxyl fatty acid chain and hydrophobic amino acid ring 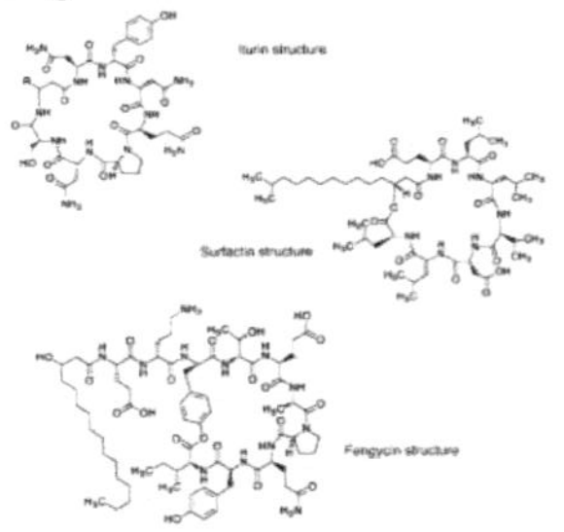	**MEOR** Soil bioremediation	Vanittanakom et al. (1986), Al-Wahaibi et al. (2014), Inès and Dhouha (2015)
Lichenysin	*Bacillus lichenformis*	Cyclic lipopeptide similar to surfactin	**MEOR**	Yakimov et al. (1995), Coronel-León et al. (2016)

(continued)

Table 3.1 (continued)

Biosurfactant type	Origin	Chemical structure	Industrial application	References
High—molecular weight biosurfactants				
Emulsan	*Acinetobacter calcoaceticus RAG-1*	A complex of a lipoheteropolysaccharide (apoemulsan) and a protein 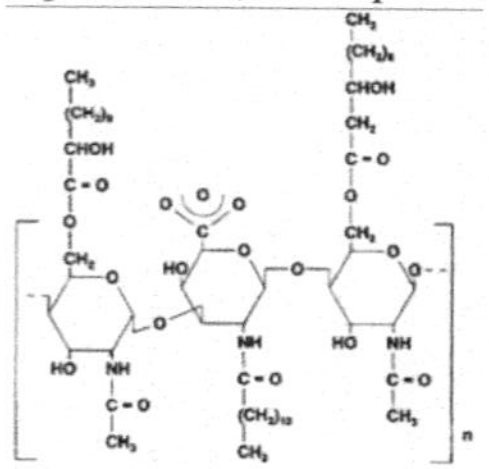	**MEOR**	Rosenberg and Ron (1999), Uzoigwe et al. (2015)
Alasan	*Acinetobacter radioresistens KA53*	A complex of anionic polysaccharides rich in alanine and proteins with high molecular mass	Bioremediation **MEOR**	Toren et al. (2001)

3.2.2 Nature and the Environment: Biosurfactant Perspective

Biosurfactant-producing bacteria are present in various habitats, including aquatic environments such as fresh and sea water, as well as groundwater, and terrestrial environments like soil, sediment, and sludge. Particular environments are known for their extreme conditions, including high/low temperature, salinities, pH, and/or pressure. These environments are also frequently inhabited by microbes that produce biosurfactants. Biosurfactants serve important ecological functions for these organisms, such as aiding in cell adhesion to surfaces and potential food sources, retaining water and concentrating nutrients, and facilitating the production of biofilms (Nicolaus et al. 2010). The production of biosurfactants by microorganisms can be directly influenced by the environment. Marine microbial EPS, as observed in studies by Gutiérrez et al. (2007) and Decho and Gutierrez (2017), exhibit elevated levels of uronic acids. This characteristic renders them polyanionic and relatively more reactive in comparison to non-marine microbial surfactants. The presence of mannose and galactosamine characterises the monosaccharides in the EPS of cold-adapted bacteria (Nichols et al. 2004). Biosurfactant-producing microbes have been isolated and cultured under laboratory conditions from extreme environments such as oil reservoirs, cold environments (e.g., polar regions), salt lakes, and hydrothermal vents. In addition, it is worth noting that hydrocarbon-polluted areas are a prime location for the search of marine biosurfactant-producing microorganisms. This is because biosurfactants are known to play a crucial role in the microbial biodegradation of hydrocarbons, as highlighted by Chandankere et al. (2014).

Many bacteria that produce biosurfactants have the ability to survive and thrive in various temperature, pH, and salinity conditions, hence displaying a diverse array of metabolic activities. Aerobic microorganisms constitute the bulk of isolated and cultured specimens due to their convenience in terms of sampling and manipulation within laboratory settings. Nevertheless, there has been a consistent increase in the discovery and effective cultivation of biosurfactant-producing microbes, predominantly consisting of anaerobic species, under ex-situ circumstances (VanFossen et al. 2008). The production of biosurfactants by anaerobic microbes has been seen in oil reservoirs, where anaerobic hydrocarbon biodegradation processes, specifically methanogenesis, take place (Nikolova and Gutierrez 2021). *Bacillus lichenformis*, as demonstrated by Al-Sayegh et al. (2015), has the capability to thrive and generate biosurfactants in both aerobic and anaerobic environments. Nevertheless, it is important to note that the biomass and surfactant production may be considerably reduced under these conditions when compared to alternative conditions, as indicated by Yakimov et al. (1995).

3.3 Employing Biosurfactants in EOR Techniques

The foaming characteristics exhibited by biosurfactants render them suitable for application in diverse industries, including the reduction of oil viscosity and the cleansing of crude oil storage tanks as detergents. Numerous researchers have conducted investigations into the utilisation of biosurfactants in enhanced oil recovery (EOR) processes. The economic feasibility of enhanced oil recovery (EOR) is hindered by the relatively higher cost of biosurfactants compared to chemical surfactants. However, biosurfactants exhibit superior performance in comparison to chemical surfactants due to their enhanced environmental compatibility, lower toxicity, and capacity to be derived from renewable sources (Kosaric 1992). By employing more cost-effective raw materials, optimising the composition of media components, increasing manufacturing efficiency, or including fermentation extracts rich in biosurfactants, it is possible to achieve a reduction in costs (Xi et al. 2021). The utilisation of biosurfactants in enhanced oil recovery (EOR) can be categorised into two main procedures: ex-situ and in-situ microbial enhanced oil recovery (MEOR), which is commonly referred to as biosurfactant-mediated MEOR (BS-MEOR) (Bryant and Lindsey 1996). In the former method, biosurfactants that are produced in a laboratory setting are introduced into the reservoir. On the other hand, the in-situ process involves the identification of suitable microorganisms already present in the reservoir, and facilitating their growth to produce necessary metabolites such as surfactants or polymers that enhance the effectiveness of enhanced oil recovery (EOR) techniques (Shahaliyan et al. 2015). After this stage, the wells are closed and subjected to monitoring in order to observe microbial activity and the generation of metabolites. MEOR field-scale in situ research efforts have yielded a notable and effective augmentation in oil output.

In their study, Lal (2009) developed a composite microbial consortium comprising three anaerobic strains that were hyperthermophilic, acidogenic, and barophilic. The objective of this consortium was to enhance oil recovery from oil reservoirs operating within a temperature range of 70–90 °C. The consortium exhibited diverse metabolic outputs, including alcohols, biosurfactants, methane, volatile fatty acids, and CO_2. These metabolic outputs were observed under controlled conditions using a specifically formulated nutrient medium. The utilisation of the microbial consortia in situ resulted in an enhanced oil recovery process, leading to improved efficiency in the displacement of crude oil from oil-saturated reservoirs within rock formations. The primary focus of research on microbial enhanced oil recovery (MEOR) applications centres around two distinct categories of biosurfactants, namely glycolipids and lipopeptides (Rassenfoss 2011). Lipopeptides have been found to exhibit enhanced effectiveness in lowering interfacial tension (IFT) and surface tension (ST) according to studies (Rassenfoss 2011; Mulligan 2005).

The large-scale synthesis of surfactin by Bacillus subtilis was investigated by Cooper et al. (1981). It was determined that surfactin could be extracted from collapsed foam via acid precipitation. The utilisation of glucose as a substrate for fermentation, along with the removal of products using foam fractionation in a sequential manner, led to a favourable yield. Furthermore, the addition of iron or manganese salts contributed to an increase in this yield. The addition of hydrocarbon to the medium resulted in a total cessation of surfactin formation by B. subtilis, despite the conventional enhancement of biosurfactant production. The surfactin can be precipitated from the wasted media of Bacillus subtilis culture by lowering the pH to 2.0. Subsequently, it can be further purified by extracting it with dichloromethane and reprecipitating it using acid.

A biochemical analysis was conducted on one of the JF-2 proteins, revealing a comparable amino acid composition and infrared spectrum to surfactin. However, it was shown that the JF-2 acid precipitate contains additional components that potentially play a crucial role in facilitating interactions among all the components and enabling JF-2 to function as a surfactant with full activity. Subsequently, solvents that are necessary for achieving full activity and interaction among these compounds have been produced. In the crude extract, four components were identified, namely chloroform, dichloromethane, and methanol. These components were shown to have sequential extraction capabilities, with chloroform extracting one component, dichloromethane extracting two components, and methanol extracting four components. The activity recovery values for these extractions were determined to be 23%, 56%, and 94% respectively. The reconstitution of surfactant preparation exhibited a higher level of activity compared to the synthesis of individual isolated components in the conducted studies (Cooper et al. 1981; McInerney et al. 1990).

According to Cooper and Goldenberg (1987), it was found that the biosurfactant derived from B. cereus, namely the monoglyceride biosurfactant, exhibited an enhancement in the emulsifier activity of polyhexosamine produced by the same organism.

Promising qualities for enhanced oil recovery were discovered through investigations conducted by McInerney et al. (1990) and Jenneman et al. (1983) on the JF-2 surfactant. The concentrations in which NaCl was present at a minimum of 5% (w/v) yielded the lowest interfacial tension (IFT). Consequently, it can be concluded that the efficacy of the JF-2 biosurfactant is enhanced when the concentration of NaCl is elevated. Oil reserves exhibit elevated salinity levels, hence providing optimal environmental conditions conducive to the proliferation of organisms. The JF-2 biosurfactant exhibits a highly favourable performance due to its remarkably low critical micelle concentration (CMC) when used at diluted proportions. This is evidenced by the achieved interfacial tension (IFT) value, which was found to be less than 0.1 mN/m. Elevated concentrations of biosurfactants or the inclusion of cosurfactants can lead to a reduction in interfacial tension (IFT).

The study conducted by McInerney et al. (1990) investigated the formation of anaerobic biosurfactants from Bacillus licheniformis strain JF-2. The bacterium was cultivated in a growth medium comprising glucose-mineral salts, $NaNO_3$, and yeast extract. The pH was reduced to 2.0, causing the precipitation of an anionic biosurfactant from the spent medium. This led to a significant decrease in the surface tension (ST) of the medium, dropping from an initial value of 70 to 74 mN/m to as low as 28 mN/m. Both JF-2 and surfactin biosurfactants exhibited a reduction in water surface tension of 27 mN/m, which represents the lowest recorded surface tension for biosurfactants. The addition of hydrocarbon to the mechanism, which typically enhances biosurfactant production, resulted in a complete cessation of surfactin formation by B. subtilis. The surfactin can be precipitated from the wasted media of Bacillus subtilis culture by lowering the pH to 2.0. Subsequently, it can be purified using dichloromethane extraction and reprecipitation using acid.

In their study, Arima et al. (1968) conducted a characterization of surfactin, a very effective biosurfactant produced by Bacillus strains. The researchers observed that surfactin exhibits significant surface activity and a critical micelle concentration (CMC) that is comparable to that of synthetic surfactants. The lipopeptide surfactin demonstrates the ability to hinder the development of fibrin clots, while also causing a decrease in surface tension and interfacial tension when exposed to hexadecane concentrations below 1 mN/m. The substance exhibits stability when exposed to elevated pH levels, increased salinity, and elevated temperatures. In addition, the utilisation of low CMC features in enhanced oil recovery (EOR) applications may confer advantages by influencing the economic aspects of EOR development (Shaligram and Singhal 2010; Banat et al. 2010).

In their study, McInerney et al. (1990) conducted the cultivation and synthesis of biosurfactants utilising the anaerobic bacterium Bacillus mojavensis JF-2, in conjunction with other coexisting microorganisms, within a sand-based ecosystem. Bacillus mojavensis JF-2 had the highest growth rate when supplied with glucose as the carbon source. Additionally, it displayed a general preference for monosaccharides as carbon sources. The fructose treatment exhibited the highest rate of yield growth. The synthesis of biosurfactants was shown to be boosted in the presence of peptone #2 (PP2) under anaerobic circumstances.

The microbial formulation comprises of four specific bacteria, known as NIPER Bac 1, which have been introduced into four designated wells located in the Delaware-Childers field in Oklahoma. The primary outcome of this introduction is the production of surfactants, alcohols, and acids. Additionally, frequent injections of molasses are administered as a nutrient source (Bryant et al. 1990).

In their study, Bryant et al. (1990) conducted experiments using micromodels to investigate the potential mobilisation of oil through microbial formulations inside a porous environment. They also examined the correlation between oil mobilisation and the effectiveness of oil recovery in Berea sandstone cores. Following the micromodel waterflooding experiment, approximately 60% of the oil has been effectively mobilised. During the incubation period of the micromodel, the occurrence of gas bubbles and the emulsification of crude oil were detected. The primary mechanisms employed by NIPER Bac 1 for oil mobilisation include gas generation, surfactants, and solvents. The preliminary findings demonstrated a notable 13% enhancement in the sustainable retrieval of oil, accompanied by a reduction in the surface tension of the extracted brine. These outcomes provide evidence for the efficacy of in situ microbial biosurfactant production as a means to augment petroleum recovery. To establish unequivocal evidence of the growth and surfactant production capabilities of strain JF-2 within an oil reservoir, the aforementioned strain was isolated from the produced brine precisely 30 weeks subsequent to its injection.

In a separate investigation pertaining to micromodels, Bryant and Douglas (1988) observed a correlation between microorganisms exhibiting a high recovery efficiency during core flooding experiments and their enhanced efficacy in mobilising crude oil. Microbial growth and the subsequent production of carbon monoxide and compounds have the potential to greatly augment oil recovery in reservoirs, provided that the salinity and temperature conditions are appropriate. The findings of a microbiological investigation conducted on Berea sandstone cores indicate significant variability in the capacity of bacteria to extract residual oil following waterflooding, with recovery rates ranging from 7.5 to 71%. Interestingly, the type of core encapsulation employed did not demonstrate any discernible impact on the quantity of oil recovered. Research conducted using a broad range of permeability values (ranging from 134 to 1,920) has demonstrated that certain microbes exhibit enhanced efficacy in the extraction of oil from cores with lower permeability. Consequently, certain bacteria that create surfactants have the ability to boost oil recovery in manners that are not achievable by microorganisms that produce gas. Therefore, it is evident that gas generation is not the sole influencing factor. It is imperative to take into account the type of surfactant, as two microorganisms have the ability to create surfactants that exhibit varying recovery capacities. Enhancing the areal sweep efficiency (EA) of microorganisms is perceived as an advantageous technique in the microbial mobilisation of oil. The comparison of recovery yields between heavy oil and medium-to-light-gravity oil indicates that the Microbial Enhanced Oil Recovery (MEOR) technology could potentially be suitable for both types of oils.

The research conducted by McInerney et al. (1990; 2002) and other studies (Thomas et al. 1993; Zhang et al. 2016) shown that the utilisation of lipopeptides resulted in an interfacial tension between water and oil that was below 0.1 mN/m. Lipopeptide surfactants exhibit low productivity and significant expenses during fermentation-based production, resulting in their infrequent use in large-scale enhanced oil recovery (EOR) initiatives. Consequently, their application is primarily confined to certain sectors such as paper, healthcare, pulp, and food industries (Hadia et al. 2019).

The work conducted by Long et al. (2017) focused on investigating the emulsifying activity of surfactin, governed by pH, and its potential application in the separation of oil in enhanced oil recovery (EOR). The findings of the study indicate that surfactin exhibits notable emulsion stabilisation properties at pH 7.4 and above. This characteristic enables convenient oil separation through pH adjustment when surfactin is employed as an emulsifying agent. The surfactin-based enhanced oil recovery (EOR) approach exhibits significant potential for use due to its ability to be utilised repeatedly and maintain its effectiveness following demulsification.

3.4 Future Perspective

Biosurfactants are considered to be highly esteemed molecules that are synthesised by microbes, and they have demonstrated significant potential for application in Microbial Enhanced Oil Recovery (MEOR). The laboratory experiments conducted on enhanced oil recovery (EOR) demonstrated a range of 10–85% increase in oil recovery. However, the field-scale research yielded a more modest additional oil recovery of approximately 3–10% over the residual oil saturation (S_{or}) (). Nonetheless, these findings remain promising. Biosurfactant-mediated Microbial Enhanced Oil Recovery (MEOR) can be implemented through two distinct approaches: in-situ and ex-situ methodologies. Numerous laboratory research and reports have demonstrated the simplicity of implementation, the reliability of data obtained from stability tests, reductions in surface tension/interfacial tension (ST/IFT), the ability to form emulsions, and the exceptional results in terms of enhanced oil recovery (EOR) percentage achieved by the ex-situ biostimulation-microbial enhanced oil recovery (BS-MEOR) strategy. Nevertheless, it is widely regarded as a theoretical construct and has yet to be effectively implemented on a large-scale basis. There are several challenges that hinder the extensive utilisation of this technology. These challenges include the significant expenses associated with manufacturing, the capacity to scale up in bioreactors, lower yields compared to chemical alternatives, and limited availability due to a relatively small number of enterprises engaged in large-scale production. The expenses associated with extraction and purification constitute a significant portion, exceeding 60%, of the overall expenditure for fermentation products, such as biosurfactants. Several potential strategies can be implemented in order to mitigate expenses. There are three strategies that can be employed to enhance the efficiency and cost-effectiveness of bacterial growth (Wu et al. 2022). Firstly, it is advisable to utilise readily available and inexpensive agro residue or industrial waste as a substrate for bacterial growth. Secondly, a combination of biological and chemical products can be utilised to optimise bacterial development. Lastly, if feasible, the use of crude microbial products should be prioritised to avoid the expenses associated with costly purification processes.

One additional concern is to the intricate reservoir environment (Wu et al. 2022), which exerts a significant influence on microbial proliferation and the efficacy of microbial displacement techniques. Hence, prior to implementing microbial flooding, it is imperative to develop a more methodical set of criteria for reservoir screening, conduct a systematic analysis of the reservoir features and microbial diversity in each well, and construct an appropriate mathematical model for Microbial Enhanced Oil Recovery (MEOR). Simultaneously, it is also possible to develop strains with exceptional resilience to challenging reservoir environments by the utilisation of genetic engineering techniques. Furthermore, the use of genetic engineering technology and the development of microbial consortium construction technology hold promising potential as a transformative approach for enhancing microbial oil recovery in the foreseeable future. The former methodology facilitates the development of oil recovery bacteria that exhibit exceptional performance, whereas the later approach yields highly efficient microbial consortia. The integration of both factors has the potential to augment the applicability of Microbial Enhanced Oil Recovery (MEOR) technology.

3.5 Conclusion

Despite being more costly than chemical surfactants, biosurfactants are favoured for their environmental compatibility, less toxicity, and capacity to be derived from renewable sources. The cost can be reduced through the utilisation of more affordable raw materials, the optimisation of media components, the implementation of hyper-production techniques, or the incorporation of fermentation extracts that possess substantial quantities of biosurfactants. The enhancement of oil recovery can be achieved through the utilisation of both in situ and ex situ microorganisms. The application of in situ microbial enhanced oil recovery (EOR) techniques involving the utilisation of metabolic products, such as biosurfactants, alcohols, methane, volatile fatty acids, and CO_2, has demonstrated enhanced oil recovery and better sweep efficiency. The investigation focused on examining the impact of several characteristics, such as pH, concentration, and components, on the behaviour of diverse types of Bacillus bacteria. The objective of this study was to investigate the effectiveness of surfactin, a biosurfactant with a low critical micelle concentration (CMC), in reducing interfacial tension. The maximisation of oil recovery efficiency was achieved through the optimisation of microbial mobilisation inside the porous environment.

References

Ahmadi MA, Arabsahebi Y, Shadizadeh SR, Behbahani SS (2014) Preliminary evaluation of mulberry leaf-derived surfactant on interfacial tension in an oil-aqueous system: EOR application. Fuel 117:749–755. https://doi.org/10.1016/j.fuel.2013.08.081

Al-Sayegh A, Al-Wahaibi Y, Al-Bahry S, Elshafie A, Al-Bemani A, Joshi S (2015) Microbial enhanced heavy crude oil recovery through biodegradation using bacterial isolates from an Omani oil field. Microb Cell Fact 14:141. https://doi.org/10.1186/s12934-015-0330-5

Al-Sulaimani H, Joshi A-W, Al-Bahry S, Elshafie A, Al-Bemani A (2011) Microbial biotechnology for enhancing oil recovery: current developments and future prospects. Biotechnol Bioinf Bioeng 1(2):147–158

Al-Wahaibi Y, Joshi S, Al-Bahry S, Elshafie A, Al-Bemani A, Shibulal B (2014) Biosurfactant production by Bacillus subtilis B30 and its application in enhancing oil recovery. Colloids Surf B Biointerfaces 114:324–333. https://doi.org/10.1016/j.colsurfb.2013.09.022

Ardabili S, Mosavi A, Várkonyi-Kóczy AR (2019) Systematic review of deep learning and machine learning models in biofuels research. In: International conference on global research and education. Springer International Publishing, Cham, pp 19–32. https://doi.org/10.1007/978-3-030-36841-8_2

Arima K, Kakinuma A, Tamura G (1968) Surfactin, a crystalline peptide lipid surfactant produced by Bacillus subtilis: isolation, characterization and its inhibition of fibrin clot formation. Biochem Biophys Res Commun 31(3):488–494. https://doi.org/10.1016/0006-291X(68)90503-2

Bachari Z, Isari AA, Mahmoudi H, Moradi S, Mahvelati E H (2019) Application of natural surfactants for enhanced oil recovery–critical review. IOP Conf Ser Earth Environ Sci 221:012039. IOP Publishing. https://doi.org/10.1088/1755-1315/221/1/012039

Banat IM (1995) Biosurfactants production and possible uses in microbial enhanced oil recovery and oil pollution remediation: a review. Bioresour Technol 51:1–12. https://doi.org/10.1016/0960-8524(94)00101-6

Banat IM, Franzetti A, Gandolfi I, Bestetti G, Martinotti MG, Fracchia L, Marchant R et al (2010) Microbial biosurfactants production, applications and future potential. Appl Microbiol Biotechnol 87:427–444. https://doi.org/10.1007/s00253-010-2589-0

Bera A, Mandal A, Guha BB (2014) Synergistic effect of surfactant and salt mixture on interfacial tension reduction between crude oil and water in enhanced oil recovery. J Chem Eng Data 59(1):89–96. https://doi.org/10.1021/je400850c

Bera A, Belhaj H (2016) Ionic liquids as alternatives of surfactants in enhanced oil recovery—a state-of-the-art review. J Mol Liq 224:177–188. https://doi.org/10.1016/j.molliq.2016.09.105

Bryant RS, Burchfield TE, Dennis DM, Hitzman DO (1990) Microbial-enhanced waterflooding: mink unit project. SPE Reserv Eng 5(01):9–13. https://doi.org/10.2118/17341-PA

Bryant RS, Douglas J (1988) Evaluation of microbial systems in porous media for EOR. SPE Reserv Eng 3(02):489–495. https://doi.org/10.2118/16284-PA

Bryant RS, Lindsey RP (1996) World-wide applications of microbial technology for improving oil recovery. In: SPE improved oil recovery conference?, pp SPE-35356. SPE. https://doi.org/10.2118/35356-MS

Chandankere R, Yao J, Cai M, Masakorala K, Jain AK, Choi MMF (2014) Properties and characterization of biosurfactant in crude oil biodegradation by bacterium Bacillus methylotrophicus USTBa. Fuel 122:140–148. https://doi.org/10.1016/j.fuel.2014.01.023

Chau TT (2009) A review of techniques for measurement of contact angles and their applicability on mineral surfaces. Miner Eng 22(3):213–219. https://doi.org/10.1016/j.mineng.2008.07.009

Chen WC, Juang RS, Wei YH (2015) Applications of a lipopeptide biosurfactant, surfactin, produced by microorganisms. Biochem Eng J 103:158–169. https://doi.org/10.1016/j.bej.2015.07.009

Cooper DG, Macdonald CR, Duff SJB, Kosaric N (1981) Enhanced production of surfactin from *Bacillus subtilis* by continuous product removal and metal cation additions. Appl Environ Microbiol 42(3):408–412. https://doi.org/10.1128/aem.42.3.408-412.1981

Cooper DG, Goldenberg BG (1987) Surface-active agents from two *Bacillus* species. Appl Environ Microbiol 53(2):224–229. https://doi.org/10.1128/aem.53.2.224-229.1987

Coronel-León J, Marqués AM, Bastida J, Manresa A (2016) Optimizing the production of the biosurfactant lichenysin and its application in biofilm control. J Appl Microbiol 120:99–111. https://doi.org/10.1111/jam.12992

Decho AW, Gutierrez T (2017) Microbial extracellular polymeric substances (EPSs) in ocean systems. Front Microbiol 8:922. https://doi.org/10.3389/fmicb.2017.00922

Desai JD, Banat IM (1997) Microbial production of surfactants and their commercial potential. Microbiol Mol Biol Rev 61(1):47–64. https://doi.org/10.1128/mmbr.61.1.47-64.1997

Geetha SJ, Banat IM, Joshi SJ (2018a) Biosurfactants: production and potential applications in microbial enhanced oil recovery (MEOR). Biocatal Agric Biotechnol 14:23–32. https://doi.org/10.1016/j.bcab.2018.01.010

Geetha SJ, Banat IM, Joshi SJ (2018b) Biosurfactants: production and potential applications in microbial enhanced oil recovery (MEOR). Biocatal Agric Biotechnol 14:23–32

Gutiérrez T, Mulloy B, Black K, Green DH (2007) Glycoprotein emulsifiers from two marine *Halomonas* species: chemical and physical characterization. J Appl Microbiol 103:1716–1727. https://doi.org/10.1111/j.1365-2672.2007.03407.x

Hadia NJ, Ottenheim C, Li S, Hua NQ, Stubbs LP, Lau HC (2019) Experimental investigation of biosurfactant mixtures of surfactin produced by Bacillus Subtilis for EOR application. Fuel 251:789–799. https://doi.org/10.1016/j.fuel.2019.03.111

Holmberg K, Shah DO, Schwuger MJ (2002) Handbook of applied surface and colloid chemistry, vol 1. Wiley Blackwell, Hoboken, NJ, USA

Inès M, Dhouha G (2015) Lipopeptide surfactants: production, recovery and pore forming capacity. Peptides 71:100–112. https://doi.org/10.1016/j.peptides.2015.07.006

Jenneman GE, McInerney MJ, Knapp RM, Clark JB, Feero JM, Revus DE, Menzie DE (1983) Halotolerant, biosurfactant-producing *Bacillus species* potentially useful for enhanced oil recovery. Dev Ind Microbiol (United States) 24(CONF-8208164)

Joshi-Navare K, Khanvilkar P, Prabhune A (2013) Jatropha oil derived sophorolipids: production and characterization as laundry detergent additive. Biochem Res Int:Article ID 169797. https://doi.org/10.1155/2013/169797

Kokal S, Al-Kaabi A (2010) Enhanced oil recovery: challenges & opportunities. World Petrol Counc off Publ 64:64–69

Kosaric N (1992) Biosurfactants in industry. Pure Appl Chem 64(11):1731–1737. https://doi.org/10.1351/pac199264111731

Lal B (2009) Process for enhanced recovery of crude oil from oil wells using novel microbial consortium. World Intellectual Property Organization, Geneva, Switzerland, Google Patents

Lake LW (2014) Fundamentals of enhanced oil recovery, vol 1. Society of Petroleum Engineers, Richardson, TX, USA

Long X, He N, He Y, Jiang J, Wu T (2017) Biosurfactant surfactin with pH-regulated emulsification activity for efficient oil separation when used as emulsifier. Bioresour Technol 241:200–206. https://doi.org/10.1016/j.biortech.2017.05.120

Makkar RS, Rockne KJ (2003) Comparison of synthetic surfactants and biosurfactants in enhancing biodegradation of polycyclic aromatic hydrocarbons. Environ Toxicol Chem 22:2280–2292. https://doi.org/10.1897/02-472

McInerney MJ, Javaheri M, Nagle DP Jr (1990) Properties of the biosurfactant produced by *Bacillus licheniformis* strain JF-2. J Ind Microbiol Biotechnol 5(2–3):95–101. https://doi.org/10.1007/BF01573858

McInerney MJ, Marsh TL, Zhang X, Knapp RM, Sharma PK, Jackson BE (2002) Development of more effective biosurfactants for enhanced oil recovery/advanced recovery concepts awards (No. DOE/BC/15113-1). National Petroleum Technology Office Tulsa, OK (US)

Mulligan CN (2005) Environmental applications for biosurfactants. Environ Pollut 133(2):183–198. https://doi.org/10.1016/j.envpol.2004.06.009

Nichols CAM, Garon S, Bowman JP, Raguénès G, Guézennec J (2004) Production of exopolysaccharides by Antarctic marine bacterial isolates. J Appl Microbiol 96:1057–1066. https://doi.org/10.1111/j.1365-2672.2004.02216.x

Nicolaus B, Kambourova M, Oner ET (2010) Exopolysaccharides from extremophiles: from fundamentals to biotechnology. Environ Technol 31:1145–1158. https://doi.org/10.1080/09593330903552094

Nikolova C, Gutierrez T (2021) Biosurfactants and their applications in the oil and gas industry: current state of knowledge and future perspectives. Front Bioeng Biotechnol 9:626639 and references there in. https://doi.org/10.3389/fbioe.2021.626639

Owens WW, Archer D (1971) The effect of rock wettability on oil-water relative permeability relationships. J Petrol Technol 23(07):873–878. https://doi.org/10.2118/3034-PA

Özdemir G, Peker S, Helvaci SS (2004) Effect of pH on the surface and interfacial behavior of rhamnolipids R1 and R2. Colloids Surf A Physicochem Eng Asp 234(1–3):135–143. https://doi.org/10.1016/j.colsurfa.2003.10.024

Park S, Lee ES, Sulaiman WRW (2015) Adsorption behaviors of surfactants for chemical flooding in enhanced oil recovery. J Ind Eng Chem 21:1239–1245. https://doi.org/10.1016/j.jiec.2014.05.040

Rassenfoss S (2011) From bacteria to barrels: microbiology having an impact on oil fields. J Pet Technol 63(11):32–38. https://doi.org/10.2118/1111-0032-JPT

Rosenberg E, Ron EZ (1999) High- and low-molecular-mass microbial surfactants. Appl Microbiol Biotechnol 52:154–162. https://doi.org/10.1007/s002530051502

Sabhapondit A, Borthakur A, Haque I (2003) Water soluble acrylamidomethyl propane sulfonate (AMPS) copolymer as an enhanced oil recovery chemical. Energy Fuels 17(3):683–688. https://doi.org/10.1021/ef010253t

Sajna KV, Sukumaran RK, Jayamurthy H, Reddy KK, Kanjilal S, Prasad RB, Pandey A (2013) Studies on biosurfactants from Pseudozyma sp. NII 08165 and their potential application as laundry detergent additives. Biochem Eng J 78:85–92. https://doi.org/10.1016/j.bej.2012.12.014

Sari CN, Hertadi R, Gozan M, Roslan AM (2019) Factors affecting the production of biosurfactants and their applications in enhanced oil recovery (EOR). a review. IOP Conf Ser Earth Environ Sci 353(1):012048. IOP Publishing. https://doi.org/10.1088/1755-1315/353/1/012048

Satpute SK, Banpurkar AG, Dhakephalkar PK, Banat IM, Chopade BA (2010) Methods for investigating biosurfactants and bioemulsifiers: a review. Crit Rev Biotechnol 30:127–144. https://doi.org/10.3109/07388550903427280

Schechter DS, Denqen Z, Orr Jr FM (1991) Capillary imbibition and gravity segregation in low IFT systems. In SPE annual technical conference and exhibition? pp SPE-22594. https://doi.org/10.2118/22594-MSPE

Schulz R (2016) The reason to expect prolonged USD 30–60/Bbl oil. J Pet Technol 68(10):42–45. https://doi.org/10.2118/1016-0042-JPT

Seethepalli A, Adibhatla B, Mohanty KK (2004) Wettability alteration during surfactant flooding of carbonate reservoirs. In SPE improved oil recovery conference?. SPE, pp SPE-89423. https://doi.org/10.2118/89423-MS

Shahaliyan F, Safahieh A, Abyar H (2015) Evaluation of emulsification index in marine bacteria *Pseudomonas* sp. and *Bacillus* sp. Arab J Sci Eng 40:1849–1854. https://doi.org/10.1007/s13369-015-1663-4

Shaligram NS, Singhal RS (2010) Surfactin–a review on biosynthesis, fermentation, purification and applications. Food Technol Biotechnol 48(2):119–134. https://hrcak.srce.hr/53621

Singh P, Patil Y, Rale V (2019) Biosurfactant production: emerging trends and promising strategies. J Appl Microbiol 126(1):2–13. https://doi.org/10.1111/jam.14057

Soñichsen N (2021) Daily demand for crude oil worldwide from 2006 to 2020, with a forecast until 2026. Statista. Accessed 22 Sept 2021

Thomas CP, Duvall ML, Robertson EP, Barrett KB, Bala GA (1993) Surfactant-based EOR mediated by naturally occurring microorganisms. SPE Reserv Eng 8(04):285–291. https://doi.org/10.2118/22844-PA

Toren A, Navon-Venezia S, Ron EZ, Rosenberg E (2001) Emulsifying activities of purified alasan proteins from *Acinetobacter radioresistens* KA53. Appl Environ Microbiol 67:1102–1106. https://doi.org/10.1128/AEM.67.3.1102-1106.2001

Uzoigwe C, Burgess JG, Ennis CJ, Rahman PKSM (2015) Bioemulsifiers are not biosurfactants and require different screening approaches. Front Microbiol 6:245. https://doi.org/10.3389/fmicb.2015.00245

VanFossen AL, Lewis DL, Nichols JD, Kelly RM (2008) Polysaccharide degradation and synthesis by extremely thermophilic anaerobes. Ann NY Acad Sci 1125:322–337. https://doi.org/10.1196/annals.1419.017

Vanittanakom N, Loeffler W, Koch U, Jung G (1986) Fengycin-a novel antifungal lipopeptide antibiotic produced by Bacillus subtilis F-29-3. J Antibiot 39:888–901. https://doi.org/10.7164/antibiotics.39.888

Wu B, Xiu J, Yu L, Huang L, Yi L, Ma Y (2022) Research advances of microbial enhanced oil recovery. Heliyon 8:e11424. https://doi.org/10.1016/j.heliyon.2022.e11424

Xi W, Ping Y, Alikhani MA (2021) A review on biosurfactant applications in the petroleum industry. Int J Chem Eng:1–10 and references therein. https://doi.org/10.1155/2021/5477185

Yakimov MM, Timmis KN, Wray V, Fredrickson HL (1995) Characterization of a new lipopeptide surfactant produced by thermotolerant and halotolerant subsurface *Bacillus licheniformis* BAS50. Appl Environ Microbiol 61:1706–1713. https://doi.org/10.1128/AEM.61.5.1706-1713.1995

Zhang J, Xue Q, Gao H, Lai H, Wang P (2016) Production of lipopeptide biosurfactants by *Bacillus atrophaeus* 5–2a and their potential use in microbial enhanced oil recovery. Microb Cell Fact 15(1):1–11. https://doi.org/10.1186/s12934-016-0574-8

Chapter 4
Insights of Potential Paraffin Wax Degrading Microbiome for Sustainable Petroleum Hydrocarbon Remediation

Sumana Pal, Subhra Satahrada, Uday Suryakanta, Swatilekha Das, Dindyal Mandal, and Ritesh Pattnaik

Abstract As a complex mixture of hydrocarbons and an essential constituent of crude oil, paraffin wax occurs as solid white form with potential applications. The rapid industrialization has witnessed global demand for crude oil since last few decades. In response to continuous rising demand, exploitation of more heavy crude oil reservoirs is being done, raising new technological issues from extraction, transportation and refinement level. However, production barrier may be critical where accumulation of paraffin wax on tubing surfaces during the extraction of oil and gas may occur. The prevention of wax accumulation frequently necessitates the usage of costly and hazardous chemical substances. To address such issue, wax-degrading bacteria may ne biotechnologically exploited. These microorganisms create microbial surfactants that effectively emulsify waxy compounds and promote the breakdown of such substances. Microbiome belonging to various genera like *Pseudomonas, Alcaligenes, Micrococcus, Nocardia, Arthrobacter, Bacillus, Rhodococcus* and *Proteus* are abundant in the environment. Such groups of bacteria possess the ability to break down residues with a high content of waxy substances. The present chapter outlines the possible microbial diversity, their function and biological mechanisms employed by paraffin wax degrading microorganisms for a sustainable remediation in petroleum contaminated environments.

Keywords Paraffin wax · Petroleum hydrocarbons · Wax degrading bacteria · Remediation · Sustainable

S. Pal · S. Satahrada · U. Suryakanta · S. Das · D. Mandal · R. Pattnaik (✉)
School of Biotechnology, KIIT deemed to be University, Bhubaneswar, India
e-mail: rpattnaik@kiitbiotech.ac.in

S. Banerjee et al. (eds.), *Environmental Friendly Green Technologies for Improvement of Heavy Crude Oil Flow Assurance*, Environmental Science and Engineering,
https://doi.org/10.1007/978-3-031-86701-9_4

4.1 Introduction

Oil being a primary and an essential energy source is widely employed in different fields, including industries, transportation and daily human activities. However, the widespread distribution system and the overexploitation of hydrocarbons pose serious threat on the environment and health owing to their due to their toxicity and their carcinogenic and mutagenic properties. Hydrocarbon pollution may possibly occur through several sources that may be natural or anthropogenic. The occurrence of oil toxicity arises due to numerous anthropogenic sources including oil extraction, during treatment field, transportation fuels accidents, leakage from underground storage tanks, petrochemical industry activities and release from oil refinery sites. In order to reduce this contamination, various chemical, physical and biological treatment methods are adopted worldwide. The choice of remediation methods depends on several parameters such as the type of pollutant and its characteristics (the physicochemical nature of the pollutant and its toxicity), the properties of contaminated site (the pollution source, the nature of the site) and the type of the pollution.

4.2 Characteristics and Composition of Paraffin Wax

Paraffin wax, a hydrocarbon with the formula of C_nH_{2n+2} and a saturated straight-chain, can be found in the form of petroleum, coal, or oil shale by-products. This wax is colorless, odorless, and tasteless, and is composed of a combination of hydrocarbon molecules that range from twenty to forty carbon atoms. Due to its distinguished physical and chemical characteristics, paraffin wax is an essential raw material. However, its environmental implications have sparked interest in comprehending its composition and behavior. Primarily made up of straight-chain and branched alkanes, paraffin wax typically contains carbon atoms ranging from C_{20} to C_{40}. The physical properties of the wax, such as its melting point, hardness, solubility and density, are determined by the length and configuration of these alkane chains. The solid-state properties of paraffin wax are a result of its crystalline structure, characterized by long alkane chains that are neatly arranged. Furthermore, paraffin wax is used as a matrix material for various composite products, which enhances their mechanical and thermal properties. Its versatile usage includes lubrication, electrical insulation, and candle making. Additionally, dyed paraffin wax can be transformed into crayons. Isoparaffin, cyclo paraffin, and naphtheno-aromatic hydrocarbons are among the hydrocarbon compounds that are typically present in paraffin. It is noteworthy to mention that solid paraffin, which exhibits a melting point ranging from 5 to 70 °C, is predominantly acquired via the distillation of petroleum that is paraffin-based in nature. Paraffin wax is categorized into various types, primarily distinguished by its composition and traits. One method of classification relies on the existence of aliphatic hydrocarbons (referred to as paraffin), wax, and high-quality oils, commonly present in paraffinic crude oils (Hsu et al. 2019). Another approach to categorization is based on the dispersion of molecular weights. For instance, paraffin waxes falling under this classification encompass a wide array of carbon numbers. Notably, those with a carbon number distribution spanning at least 18 carbon atoms and covering around 95% of the composition are utilized for diverse purposes. Furthermore, the formulation of a waxy mixture can be achieved by blending paraffin using both the paraffin 90 program and the paraffin 60 bp program. This blend, along with a small quantity of essential oils, becomes a resource for crafting flavored items and ornamental pieces. Moreover, the construction of a paraffin wax concoction is attainable by melding low-melting-point paraffin, high-melting-point paraffin, and microcrystalline wax. This resultant combination serves as a stabilizing element within dispersion polymerization reactions. In the final analysis, a paraffin wax mixture can be composed of a pliant wax compound containing isoparaffins, in combination with elastomeric block copolymers and/or microcrystalline wax. This mixture also incorporates a content of oil, ranging from 5 to 35% by weight, and features the inclusion of elastomeric block copolymers and/or microcrystalline wax (Andre et al. 2004). This formulation of paraffin wax encompasses constituents like low-melting-point paraffin, high-melting-point paraffin, and microcrystalline wax, with a composition comprised of n-alkanes that constitute 60–90% by weight. Moreover, this mixture displays a standard distribution of carbon numbers. Notably, paraffin waxes characterized by a broad range of molecular weights exhibit enhanced attributes, such as heightened hardness, well-suited blend kinematic viscosity, and superior capabilities in terms of wet strength and waterproofing (Agrawal and Hoch 2016). The components and factors influencing the qualities of a paraffin wax emulsion encompass a range of considerations, including the composition of the emulsifier, its quantity, the temperature, rotational speed, and the duration of the emulsification process (Feng et al. 2011).

Paraffin wax is constituted by the combination of paraffin wax itself with either alkenyl succinic acid or anhydride, leading to an enhancement in its hardness. A key constituent within high-molecular compounded and modified wax is paraffin wax, making up a significant proportion (ranging from 85 to 92 parts by weight), in conjunction with the inclusion of polyvinyl wax. The nature of paraffin wax lies in its formulation as a blend of linear hydrocarbons, resulting in a mixture with varying melting points (Allison 1998). The composition of paraffin involves a content ranging from 25 to 55% by weight of aluminum, aluminum oxides/silicates, magnesium, or magnesium oxides/silicates, leading to an enhancement in its heat storage capabilities. Paraffin-based compositions are predominantly composed of n-eicosane, accompanied by smaller quantities of n-octadecane and n-nonadecane. These compositions possess a melting point lower than that of pure n-eicosane (Krasavtsev et al. 2017). When determining the melting points of blends containing paraffin and ceresine, it is possible to employ an additive principle, but only if the melting points of the individual components differ by no more than 15 °C. For coating applications, paraffin wax compositions with a wide-ranging distribution of molecular weights have been developed. These formulations offer improved characteristics, including heightened congealing points, as well as enhanced wet strength and water resistance. The suitability of beeswax, candelilla wax, and paraffin wax as coatings for cheeses has been investigated, with beeswax and combinations of beeswax along with candelilla wax or paraffin wax demonstrating the most favorable outcomes. Enhanced attributes for coating applications, including corrugated cardboard, have been observed in paraffin wax formulations with a wide-ranging distribution of molecular weights. These improvements encompass heightened wet strength, enhanced water resistance, and elevated temperature durability. In the realm of melting points, it is observed that paraffin-ceresine blends conform to the additivity principle. To be more specific, paraffins, soft paraffins, and waxes demonstrate crystallization in the H and R phases, whereas ceresines exhibit crystallization solely in the R phase. A waxy concoction tailored for flavoring and ornamental items involves a fusion of the paraffin 90 program and the paraffin 60 bp program. This amalgam exhibits varying hardness and viscosity characteristics and is augmented with essential oils at a temperature of approximately 70 °C. Paraffin wax compositions characterized by a wide distribution of molecular weights deliver heightened qualities in the realm of coating applications, notably for corrugated cardboard. These attributes encompass improved wet strength, enhanced waterproofing, and enhanced resistance to high temperatures (Agrawal and Hoch 2019).

Paraffin waxes solubility is notably influenced by temperature changes. Factors that result in a reduction of oil temperature contribute to the process of wax crystallization. Typically, paraffin waxes remain dissolved within crude oil, maintaining an equilibrium state, especially in standard reservoir conditions. However, when this equilibrium is disturbed by temperature and pressure fluctuations, paraffin can undergo crystallization or precipitation. Another scenario that can prompt paraffin precipitation is the absence of volatile light constituents, which naturally serve as solvents (Misra et al. 1995). Paraffin wax has been examined for its physical and mechanical attributes, encompassing crystal phases, absorption bands, and melting temperature. Its characteristics include a substantial regression rate and commendable mechanical strength. Nonetheless, the drawback of paraffin wax lies in its limited thermal conductivity, which curtails its viability in temperature-control applications. Combining paraffin wax with hydroxyl-terminated polybutadiene (HTPB) can enhance the mechanical properties of the latter. Additionally, paraffin wax finds utility as a phase change substance within thermal systems, with investigations into its melting behavior within a rotating rectangular enclosure (Farsani et al. 2020). The rotational aspect of the enclosure can influence the rate of melting and the transfer of heat associated with the paraffin wax.

4.2.1 Significance of Paraffin Wax Degradation in Bioremediation

The escalating scope of industrial operations and human-driven interventions has resulted in the pollution of natural environments with hydrocarbon pollutants, including paraffin wax. Paraffin wax, a complex amalgamation of saturated hydrocarbons, presents a notable environmental predicament due to its stubborn persistence and resistance to breakdown. Nevertheless, recent strides in bioremediation methodologies have highlighted the promise of microbial communities in facilitating the breakdown of paraffin wax. The significance of paraffin wax degradation through bioremediation is multifaceted. Primarily, it offers an efficient avenue for eliminating paraffin wax from oil wells and pipelines, rendering it both environmentally sound and cost-effective (Patel and Bhaskaran 2018). The challenges posed by conventional techniques for wax removal (mechanical, thermal, and chemical), the emergence of microbial approaches has offered a practical alternative. Microbial remediation encompasses the utilization of microorganisms or their metabolic byproducts, such as surfactants and paraffin solvents, to effectively address and mitigate issues associated with paraffin. Notably, Rana et al. have recently introduced innovative systems involving bacterial consortia, fortified by the inclusion of nutrient supplements. These systems present a novel method to counteract the accumulation of paraffin in both oil wells and surface flow lines. Microorganisms, particularly hydrocarbon-degrading bacteria and fungi, play a pivotal role in initiating and sustaining the process of paraffin wax degradation. Notably, strains such as *Pseudomonas* sp. PW-1 and *Geobacillus kaustophilus* TERI NSM have showcased the capability to degrade paraffin wax, positioning them as potential candidates for addressing paraffin deposition issues in oil wells. *Nocardia farcinica* has demonstrated significant efficacy in the degradation of the fundamental components of paraffin wax, indicating its potential in enhancing the flow velocity of unrefined petroleum via pipelines. Additionally, the utilization of microbial remediation, employing bacteria and fungi, offers a substitute to traditional mechanical and chemical methodologies for paraffin elimination. Moreover, certain bacterial strains, such as *Rhodococcus erythropolis* and *Pseudomonas* sp., have demonstrated the capacity to decompose paraffin wax and employ it as a source of carbon (Rodrigues and Carvalho 2019).

The degradation of paraffin wax in crude oil has primarily been attributed to mesophilic bacteria, including *Bacillus cereus* QAU68 and *Bacillus subtilis* XCCX; *Pseudomonas aeruginosa* B1 and *Acinetobacter junii* B2; and *Rhodococcus* sp. Moj-3449. However, considering that the operational conditions within pipelines and petroleum reservoirs may occasionally exceed 50 °C, the role of thermophilic hydrocarbon-degrading microorganisms becomes pivotal. Thermophilic bacteria exhibit the unique capability to endure elevated temperatures while effectively breaking down crude oil. Among the thermophilic bacteria with reported crude oil degradation potential are *Geobacillus kaustophilus* TERI NSM, *Geobacillus pallidus* H9, and *Geobacillus* sp. ZY-10. Although some previous studies have explored thermophilic *Geobacillus* spp., certain instances required longer incubation periods such as eight, ten, or twenty-two days. Moreover, it is noteworthy that many investigations concerning *Geobacillus* spp.'s role in oil degradation have focused on a narrower range of hydrocarbon decomposition (Adlan et al. 2020a).

These microorganisms have the capacity to create biofilms and adapt their cellular membrane composition, which in turn enhances the biodegradation of paraffin. Moreover, the degradation procedure generates gases and organic acids, which contribute to the further breakdown of paraffin wax. Collectively, the bioremediation-driven degradation of paraffin wax presents a promising remedy for tackling paraffin accumulation challenges within oil wells and augmenting oil retrieval. The importance of paraffin wax degradation within the scope of bioremediation extends to the conversion of hazardous compounds into less harmful substances, along with the potential to counter hydrocarbon pollutants in the environment (Hao et al. 2008). This significance is underscored by its role in averting the buildup of wax crystals on borehole surfaces and elevating oil production. The essence of paraffin wax degradation in bioremediation lies in its ability to effectively diminish the viscosity of raw petroleum, thereby preventing the accretion of paraffin crystals on well walls. Additionally, this degradation process serves as a carbon source for microorganisms engaged in biodegradation activities. In a broader context, the bioremediation-induced breakdown of paraffin wax substantially ameliorates the fluidity of crude oil by reducing viscosity and eliminating wax constituents (Wang et al. 2019). The process contributes to the reduction of crude oil viscosity, enhancement of fluidity, and efficient extraction of wax from oil wells. This microbial process of degradation entails the production of enzymes that enhance the solubility of paraffin wax while breaking down its constituent elements. Gaining insights into the mechanisms governing the degradation of paraffin wax holds the potential to craft strategies for its elimination and enhance oil retrieval.

Furthermore, in the context of horse racing, the breakdown of wax-based binders used in synthetic track surfaces can impact track behaviour. The analysis of these alterations can provide valuable guidance for decisions regarding maintenance and replacement. The biodegradation of paraffin wax entails intricate metabolic pathways orchestrated by a diverse array of enzymes, which include alkane hydroxylases, monooxygenases, dioxygenases, alcohol dehydrogenase, lipase, and esterase. These enzymes have a role, in aiding the conversion of molecules found in paraffin wax into intermediary forms that can be assimilated by microbial cells. The effectiveness of paraffin wax degradation in bioremediation processes is influenced by various factors. Environmental parameters, including temperature, pH, nutrient availability, and oxygen levels, exert a significant influence on microbial activity and thereby influence the degradation rate. Moreover, the composition of paraffin wax, the coexistence of other pollutants, and the adaptability of microbial communities collectively contribute to the outcomes and efficacy of bioremediation strategies.

Bioremediation refers to the utilization of organisms to eliminate or neutralize pollutants from contaminated sites, constituting a form of waste management. This technique primarily involves the transformation of hazardous substances into less harmful or non-toxic forms through the actions of organisms. Bioremediation encompasses diverse technological approaches. One such approach involves two categories: in situ and ex situ. In the in situ method, the treatment of contaminated soil or water occurs directly at the location of contamination, while the ex situ method involves transporting the contaminated materials to an external location for treatment (Raju and Scalvenzi 2018). By harnessing the inherent metabolic capabilities of microorganisms, bioremediation offers a promising route to deconstruct paraffin wax into simpler compounds. The breakdown of paraffin wax via biodegradation encompasses processes like biofilm development, bio-surfactant generation, and phenotypic adjustments in bacterial cells. Collectively, the bioremediation-driven degradation of paraffin wax presents a hopeful avenue for effectively addressing environmental pollutants.

4.3 Microbial Diversity Involved in Paraffin Wax Biodegradation

In oil reservoirs, paraffins are in a state of balance when exposed to high temperatures and pressures. Nevertheless, when crude oil is brought to the surface through pumping, there is a notable decrease in both temperature and pressure. If the thermal conditions of the system decrease lower than the wax appearance temperature, it is feasible for the paraffin molecules present in crude oil to undergo a process of solidification and amass on the internal linings of the pipelines. The accumulation of paraffin may lead to significant implications for the production procedure, as well as the volume of crude oil that can be obtained (Xiao et al. 2012). This problem is commonly observed in both oil production and transportation systems, especially in pipelines situated in deep-sea environments. The blockage of pipelines and the resultant escalation in production costs are some of the negative outcomes of this issue. Therefore, it is critical to devise effective solutions to mitigate the effects of paraffin deposition in oil pipelines. An effective strategy to tackle this problem involves altering the oil's density by breaking down heavier molecular weight paraffin into lighter counterparts, thereby decreasing the apparent molecular weight. This method has gained notable interest due to the presence of naturally occurring microorganisms, their qualities of being non-flammable, non-cancer-causing, and non-disease-causing, as well as their environmentally friendly and economically viable attributes (Mokhatab and Towler 2009).

According to research studies several bacterial cultures possessed ability to produce bio-surfactants. These bacterial cultures were sourced from seawater sediments that had undergone contamination with oil. Extraction of the biosurfactants from the bacterial cultures was then performed. *Pseudomonas species* and *Actinomyces species* were among the strains that were discovered. These strains demonstrated elevated emulsification activity, and they were effective in reducing heavy hydrocarbon fractions while raising C_{15}–C_{20} fractions in crude oil. When treated with Pseudomonas and *Actinomyces species*, the wax appearance temperature and crude oil viscosity decreased, indicating an improvement in the crude oil's properties. The findings from the study also emphasized the potential of isolate PRCW B1 to achieve a higher level of emulsification activity and biodegradation of heavy hydrocarbons compared to isolate PRCW E1. Past studies have shown that *Pseudomonas* and *Bacilli* species are proficient in breaking down heavy hydrocarbons and preventing the formation and accumulation of wax in oil systems. A detailed information about involvement of microbial diversity is provided in Table 4.1.

The bacterial strains that were isolated were typically viable, naturally existing, capable of thriving in both aerobic and anaerobic conditions, non-harmful, and environmentally friendly. All strains displayed catalase activity and were aerobes, except for a single strain, which had the ability to function as a facultative anaerobe. The bacterial strains' environmental safety is a significant advantage for their use in the bioremediation of oil-contaminated seawater. The process of bioremediation offers an environmentally conscious and economically effective method for eradicating oil pollutants from seawater. The present findings contribute to the development of a more comprehensive understanding of the role of bacterial strains in the bioremediation of oil-contaminated seawater. The utilization of these bacterial strains in the bioremediation process will result in significant progress toward the achievement of sustainable development goals. Additionally, our results provide a foundation for future studies on the bioremediation of oil-contaminated seawater (Etoumi 2007).

The identification of a new bacterium, named *Gordonia amicalis* LH3, with the ability to break down paraffin, as reported by Hao et al., bears great importance due to its auspicious potential in diverse areas including paraffin management, microbial enhanced oil recovery (MEOR), and bioremediation of hydrocarbon-contaminated surroundings. In aerobic conditions, the bacterium can degrade approximately 18.0%

Table 4.1 Microbial diversity involved in potential bioremediation of paraffin wax

Bacterial species	Main degradation profile
Pseudomonas sp.	n-Alkanes (C_{14}–C_{30}) Cyclohexane Resins
Rhodococcus sp.	n-Alkanes (C_{13}–C_{17})
Geobacillus thermodenitrificans	n-Alkanes (C_{15}–C_{36})
Pseudomonas aeruginosa	Mono-aromatics
Pseudomonas sp. *Burkholderia* sp. *Rhodococcus* sp. *Mycobacterium* sp.	C_5–C_{16} alkanes, fatty acids, alkyl benzenes, cycloalkanes
Acinetobacter sp. *Caulobacter* sp. *Mycobacterium* sp.	C_5–C_{16} alkanes, cycloalkanes
Acinetobacter sp.	C_{10}–C_{30} alkanes
Bacillus sp. *Pseudomonas* sp.	Asphaltenes
Bacillus licheniformis	Polyaromatics
Dietzia sp.	n-Alkanes (C_6–C_{40})

of paraffin (2 wv) in ten days of incubation, with a preference for C_{18}–C_{24} alkanes. Meanwhile, under anaerobic conditions, its ability to degrade paraffin decreases to approximately 18% of that seen in aerobic conditions, with a preference for C_{34} and C_{36} alkanes. Furthermore, Gordonia amicalis LH3 can thrive on oil, resulting in a 44.7% reduction in oil viscosity and a 10.4% degradation of oil when exposed to aerobic conditions. These findings can have significant implications for the oil industry and environmental preservation efforts (Hao et al. 2008).

Zhang and colleagues have made a significant discovery of a bacterium, *Pseudomonas* sp. strain PW-1, which degrades paraffin wax and is found in petroleum-contaminated sites in Daqing, China. The identification of the strain was accomplished through the analysis of 16S rDNA sequence and biochemical and physiological characteristics. The strain has the ability to enhance oil recovery in oil wells that have problems with paraffin deposition. The optimal degradation conditions for the strain have been ascertained, and the theoretical degradation rate was calculated to be 54.68%. Experimental results have shown a degradation rate of 52.85% when using the strain as the sole carbon source. These results demonstrate that the PW-1 strain has potential applications in the oil industry for addressing paraffin deposition problems and enhancing oil recovery (Zhang et al. 2016).

4.4 Enzymes Involved and Its Mechanism

In this particular study, an assessment was conducted on the degradative enzymes, namely Alkane monooxygenase, alcohol dehydrogenase, lipase, and esterase. The study identified *Geobacillus kaustophilus* N3A7, *Geobacillus kaustophilus* DFY1, *Geobacillus jurassicus* MK7, *Geobacillus stearothermophilus* AD11, and *Geobacillus stearothermophilus* AD24 as the primary sources of these enzymes. It was observed that strains N3A7, MK7, DFY1, AD11, and AD24 had higher levels of alcohol dehydrogenase, lipase, and esterase activities as compared to other strains. Conversely, strains DFY3, AZ72, SA36, NFA23, T7, and D9 exhibited lower levels of esterase and lipase activity when compared to the proficient degrader strains. The study also found that most strains exhibited an increase in esterase and lipase activity after 2 or 3 days of incubation, with the highest induction being observed on the second or third day.

Alkane monooxygenase, alcohol dehydrogenase, lipase, and esterase are essential enzymes involved in the breakdown of paraffin wax within crude oil during the degradation process.

Initiation of alkane degradation in crude oil is a fundamental role played by alkane monooxygenase. The alteration of the structure of alkane hydrocarbons by the bacterial strains' production of extracellular enzymes facilitates the bacterial uptake of metabolic intermediates through the cell membrane. The enzymatic degradation of long-chain n-alkanes involves either breaking the large aliphatic chains apart or creating internal linkages within the substrates to transform them into short-chain n-alkanes. This process ultimately improves the degradation and uptake of

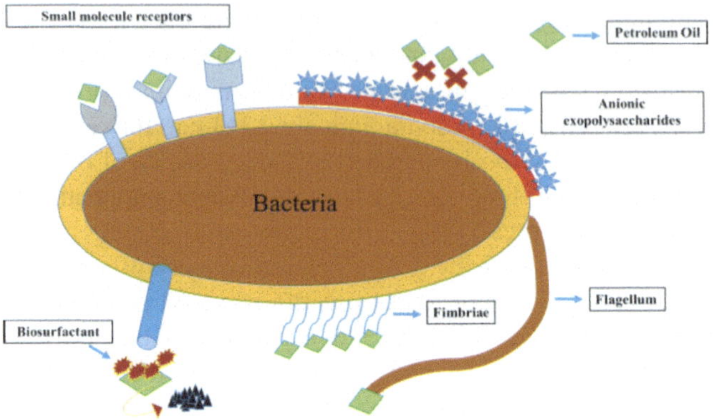

The fimbriae and flagellum of bacteria possess the ability to adhere to petroleum oil, while the bio-surfactant secreted by bacteria have the capability to emulsify this oil. Additionally, the presence of proteins, lipids, and other small molecules on the bacterial surface aids in the adhesion of petroleum oil. Moreover, certain anionic exopolysaccharides found on the bacterial surface serve as a preventive measure, inhibiting the attachment of bacteria to petroleum oil.

Fig. 4.1 Mechanism of microbially mediated remediation of paraffin wax

hydrocarbons into bacterial cells for further biodegradation. The activity of these degradative enzymes is induced by the accumulation of these products. Collectively, the enzymes examined in this study are instrumental in the breakdown of paraffin wax in crude oil, leading to enhanced degradation efficiency and the production of metabolic intermediates that can be further utilized by bacterial strains (Fig. 4.1).

4.5 Ecological Factors Influencing Paraffin Wax Biodegradation

Nutrient availability plays a significant role in influencing the rate of hydrocarbon biodegradation in marine environments. It is widely recognized that the availability of nutrients is the primary constraint in this process when it occurs at sea. To address this constraint, the experiments in this study were carried out in conditions where nutrient limitations were not an issue, mirroring the same nutrient enrichment levels found in the St. Lawrence Estuary. This approach allowed for an accurate assessment of the impact of other factors on biodegradation (Siron et al. 1995).

One determining factor in this regard is the water mass, which research has shown to be more significant than the quantity of spilled oil when it comes to regulating biodegradation. As long as there is an adequate supply of oil that is readily available for microbial consumption in the water, the process of biodegradation continues, maintaining a robust level of bacterial activity. This finding has important implications for oil spill response efforts, as it suggests that the focus should be on dispersing the oil and preventing it from accumulating in specific areas. Another factor that can impact hydrocarbon biodegradation is phytoplankton development. The presence of natural organic enrichment resulting from the spring phytoplankton bloom has been observed to enhance biodegradation in the tank with the lowest level of pollution. Changes in specific hydrocarbon ratios over time suggested that, under these circumstances, a distinct biodegradation process was at play. This underscores the significance of taking into account the unique environmental conditions in which biodegradation occurs (Lee and Levy 1989).

Temperature is another critical factor that can affect hydrocarbon biodegradation. In arctic/subarctic environments, the winter period is particularly challenging due to reduced biodegradation under icy conditions. However, biodegradation can still occur at sub-zero temperatures, albeit at lower rates. This finding has important implications for understanding the potential for biodegradation in colder regions and for developing effective response strategies for oil spills in these environments. Overall, these findings highlight the complex interplay of various factors that can impact hydrocarbon biodegradation in marine environments and underscore the importance of considering the specific environmental context in which biodegradation occurs. A detailed overview of all essential factors influencing the hydrocarbon degradation is highlighted in Table 4.2.

4.5.1 Metagenomics Approaches

Molecular techniques, such as metagenomics, have been used to study paraffin degrading biomes and understand the dynamics of microbial communities involved in the degradation process. Metagenomics analysis is a pivotal factor in comprehending and scrutinizing genomic data in the context of bioremediation experiments. This approach offers valuable insights into the operational and structural aspects of communities engaged in the degradation of paraffin compounds. By combining methods, with advanced high throughput techniques, in molecular biology, chemistry and physics researchers have gained a deeper understanding of the complex mechanisms involved in paraffin bio deterioration (Oppenheimer and Hiebert 1989). Metagenomics has been particularly useful in identifying the microbial main actors of parchment bio-deterioration and their dynamics, contributing to the preservation of ancient documents. Molecular methods present valuable techniques for identifying and examining particular microorganisms in paraffin-embedded biopsies, particularly when conventional culture-based methodologies are not viable. The overall outcome of microbial metabolic processes involves interactions that have significance on a scale. There are two types of interactions, for the completion of cycles, which are crucial for supporting life on Earth. The first type of interaction is related to microbiology, where communities of microorganisms work together as a group and achieve success collectively, than individually (Lozupone and Knight 2008). Chemical interdependence plays a role, in the interactions among members of communities. These interactions can range from being obligatory to minimal. It is widely accepted that microbial communities that interact differ from those where members

Table 4.2 Various essential factors influencing the hydrocarbon degradation in environments

Factors	Applications	Comments
Nutrient availability	Plays a significant role in hydrocarbon biodegradation	Nutrient limitations are a primary constraint in the process at sea. Experiments conducted under conditions mirroring nutrient enrichment levels in the St. Lawrence Estuary
Water mass	More significant than the quantity of spilled oil in regulating biodegradation. Adequate oil supply in the water sustains biodegradation	Focus in oil spill response should be on dispersing oil to prevent accumulation in specific areas
Phytoplankton development	Presence of natural organic enrichment from spring phytoplankton bloom enhances biodegradation	Observed impact on the tank with the lowest pollution level. Unique environmental conditions influence distinct biodegradation processes
Temperature	Critical factor affecting hydrocarbon biodegradation	Reduced biodegradation under ICY conditions in arctic/subarctic environments during winter. Biodegradation still occurs at sub-zero temperatures, albeit at lower rates
Interactive relationships	Complex interplay of various factors influences hydrocarbon biodegradation	Specific environmental conditions (e.g., nutrient levels, water mass, temperature) are crucial. Importance of considering the unique context for effective response strategies Presence of pollutants, heavy metals, or other stressors can either inhibit or stimulate microbial activity, impacting wax biodegradation
pH	Microbial activity and enzyme efficiency can be influenced by the pH of the environment	Optimal pH conditions may vary among microorganisms
Nutrient deficiency	Adequate levels of nutrients, such as nitrogen and phosphorus, are essential for microbial growth and wax degradation	

simply coexist. Around a century ago scientists began questioning whether microbial species are found everywhere or if they are limited, to regions. This inquiry arose alongside the need to understand the diversity of species worldwide. Current evidence indicates that many microorganisms are not universally distributed but restricted to habitats and geographic locations. Despite this, there are several examples of genuinely cosmopolitan organisms, such as the marine obligate hydrocarbon-degrading *Gammaproteo bacteria* and the Deep-Sea Marine Group I, specifically Alcanivorax (DeLong 2006). The advent of novel and sophisticated technologies may reveal the existence of microorganisms that were previously underestimated. A methodical investigation of this crucial variation will enable a novel comprehension of metabolic routes and the mutually reliant linkages within microbial subsistence, revealing the roles executed by distinct organisms in diverse ecological systems.

In this particular context, it is of utmost importance to ascertain the microbial and enzymatic complements across diverse niches and their consequential influence on the functioning of the community. Moreover, the existing and forthcoming approaches in systems microbiology offer a way to comprehend the intricate characteristics of microbial communities, their behavior over time, and the consequences they have on natural environments. The utilization of systems-oriented techniques towards microbial communities can potentially provide answers to the core inquiries of environmental microbiology, including the recognition of prevailing organisms and their functions. Nonetheless, the longstanding obstacle in environmental microbiology highlights that merely a negligible fraction of microorganisms can be easily cultivated. Therefore, the primary step in these studies involves identifying the constituents of the community under investigation as well as their interrelationships (Ingham et al. 2007).

4.5.2 Isolation and Cultivation of Paraffin Degrading Strains

Several studies have reported the isolation and cultivation of bacteria that possess the ability to degrade paraffin. A range of bacterial strains that possess the ability to degrade both paraffin wax and solid wax have been obtained from the body of the citrus mealy bug. Out of these isolates, four strains, namely *Bacillus cereus*, *Bacillus subtilis*, *Bacillus licheniformis*, and *Pseudomonas putida*, demonstrated impressive wax degradation capabilities. Similarly, another study isolated five paraffin-degrading bacteria from petroleum-contaminated soil, with the dominant strain being identified as SS13. Additionally, *Bacillus amyloliquefaciens* 6-2c, a bacterial strain that produces biosurfactants, was obtained from oil-contaminated soil. This strain was found to exhibit efficient paraffin degradation and enhanced oil recovery. Moreover, *Pseudomonas* sp. strain PW-1 was discovered in areas contaminated by petroleum and displayed proficiency in paraffin wax degradation, positioning it as a practical choice for use in oil wells facing challenges related to paraffin buildup (Zhang et al. 2016). The current investigations bring to notice the possibility of isolating and fostering bacterial strains that exhibit the ability to decompose paraffin, which can be utilized in a wide range of applications within the oil industry. Employing a microbial strategy with the objective of thwarting paraffin build-up in oil well tubing presents a highly encouraging prospect as it is both non-hazardous and economically feasible. It is widely recognized that alkanes, which constitute paraffins, can be effectively degraded by various microbial agents. Among these, *Pseudomonas putida* GPo1 has been extensively studied as an alkane-degrading bacteria. This specific variety is recognized for its exceptional ability to effectively employ medium and long-chain alkanes spanning from C_{12} to C_{22} in terms of carbon chain length. In addition, *Pseudomonas aeruginosa* strain WatG has been shown to degrade paraffinic alkanes such as hexatriacontane (C_{36}) and tetracontane (C_{40}) in a mineral salts medium containing crude oil. Additionally, numerous Acinetobacter strains have been recognized as effective decomposers of paraffinic hydrocarbons. Among these, strain M-1 stands out, notably for its capacity to break down paraffins spanning carbon chain lengths from C_{20} to C_{44}. Koma et al. (2001) documented an Acinetobacter strain capable of efficiently breaking down long-chain n-paraffins from automobile engine waste oil under conditions at 30 °C (Koma et al. 2001). In a separate study, Kotlar et al. accomplished the isolation of an *Acinetobacter* sp. 6A2 strain with the capability to utilize paraffins within a melting point range of 52–54 °C. Although there exist various bacteria that can degrade alkanes within mesophilic temperature ranges spanning from 30 to 37 °C, the number of reports on thermophilic bacteria that exhibit efficient wax degradation as a carbon source is limited. In 2002, Marchant et al. identified strains of *Geobacillus* that can grow at temperatures ranging from 40 to 80 °C and utilize alkanes (Marchant et al. 2002). The discovery of thermophilic bacterial strains with the ability to biodegrade n-alkanes has been limited in scope.

Nonetheless, it is noteworthy that *Bacillus stearothermophilus*, a strain found in an oil-contaminated Kuwaiti desert, has the capability to grow on C_{15}–C_{17} n-alkanes. Additionally, two strains of *G. jurassicus*, derived from a high-temperature petroleum reservoir, have been identified to thrive on C_6–C_{16} n-alkanes. Moreover, it has been demonstrated that *Bacillus thermoleovorans*, a particular strain sourced from subterranean petroleum reservoirs, exhibits a remarkable aptitude for decomposing n-alkanes of up to C_{23} at a temperature of 70 °C. In a comparable vein, an analysis has identified *Thermus brockii* as having the ability to degrade hexadecane at 70 °C. It's important to highlight that the predominant route for n-alkane degradation typically follows a pathway known as terminal oxidation. This course of action entails the initial oxidation of n-alkanes to generate the corresponding primary alcohol, which is then subjected to subsequent oxidation steps that encompass alcohol and aldehyde dehydrogenases, respectively. The plasmid OCT, found in Pseudomonas putida Gpo1, is responsible for encoding the terminal oxidation pathway that facilitates the degradation of n-alkanes. The initial stage of this pathway is catalyzed by a membrane-bound n-alkane hydroxylase system, which comprises of an integral membrane monooxygenase encoded by alkB, as well as a rubredoxin and a rubredoxin reductase. The functional evaluation of AlkB homologs has been conducted in diverse bacterial strains that possess the ability to degrade n-alkanes, and their existence has been recorded in these strains (Beilen et al. 2003). The availability of hydrophobic pollutants, such as n-alkanes, is a major limiting factor for their degradation by microorganisms. To enhance their availability, several strategies are employed, such as biofilm formation and biosurfactant production (Bognolo 1999). Moreover, it is noteworthy that elevating the temperature can lead to an improvement in the bioavailability of hydrocarbons by encouraging the proliferation of thermophilic microorganisms that are involved in hydrocarbon degradation, as previously indicated. In comparison to mesophilic organisms, some studies have highlighted the advantages of utilizing thermophilic microorganisms for hydrocarbon bioremediation. The increase in temperature can lead to an elevation in the solubility of hydrophobic contaminants, a reduction in their viscosity, an improvement in their distribution, and an acceleration in the alteration of long-chain n-alkanes from a solid to a fluid state.

4.6 Bioremediation Strategies Using Paraffin-Degrading Microbiomes

Petroleum pollution has emerged as a worldwide environmental issue, with soils tainted by petroleum hydrocarbons exerting detrimental effects on soil health. These effects occur even at lower concentrations than those impacting human well-being. The damage extends to soil microorganisms, leading to a decrease in their quantity and functionality. In addressing this concern, bioremediation has gained prominence a technique involving the use of microorganisms to break down contaminants. This method has found widespread application in treating sites contaminated with oil, offering an effective, budget-friendly, and ecologically sound approach to remediation. Within bioremediation, two primary strategies are employed: bio-augmentation and bio-stimulation. The first approach pertains to the introduction of microorganisms with the aim of augmenting degradation whereas the latter concentrates on amplifying the metabolic activity of extant microbial communities via nutrient supplementation. Nitrogen and phosphorus are indispensable nutrients that foster microorganism growth. A multitude of studies underscores the import of providing nitrogen and phosphorus supplements to intensify the disintegration of petroleum in contaminated soil (Leys et al. 2005). Based on current research findings, it can be concluded that bacterial activity during bio-stimulation can withstand exposure to a range of hydrocarbons when a C/N/P ratio of 100/5/1, 100/10/1, or 100/15/1 is maintained. Furthermore, these hydrocarbons can also serve as a viable carbon source for the growth of these bacteria. Typically, a ratio of 100/10/1 for carbon/nitrogen/phosphorus (C/N/P) is widely recognized as the standard formula in the bio-stimulation approach (Shahi et al. 2016). In cases where contaminated soils possess limited indigenous populations of hydrocarbon-degrading microorganisms, bio-augmentation comes forth as the optimal choice. Furthermore, bio-augmentation proves effective when pollutants exert a toxic impact on the native microbial community. Notably, a significant benefit of bio-augmentation lies in the expeditious commencement of the degradation process upon the introduction of particular microbial degraders. Certain research suggests that while bio-augmentation offers a temporary boost in degradation efficiency, bio-stimulation emerges as a more optimal remediation approach for achieving thorough petroleum decontamination (Kauppi et al. 2011). Within soil ecosystems, microbial communities play a crucial role and are indispensable for addressing petroleum hydrocarbon and other pollutant issues. The efficacy of petroleum hydrocarbon remediation is contingent upon the metabolic capabilities of soil microorganisms. It is therefore imperative to fully comprehend the ramifications of bio-augmentation and bio-stimulation on the composition of soil microbial communities in order to guarantee the successful remediation of oil-contaminated soil (Megharaj et al. 2011).

4.6.1 Bio-augmentation

An effective bioremediation initiative typically demands tailored strategies designed to match the unique environmental circumstances of the polluted location. Among the frequently employed choices for bio-augmentation are: introducing a pre-acclimated, pure bacterial strain; introducing a pre-acclimated mixture of microorganisms; incorporating genetically modified bacteria; and introducing relevant biodegradation genes enclosed within a vector for transfer via conjugation into native microorganisms (Fantroussi and Agathos 2005). Employing a microbial consortium rather than a pure culture for bioremediation presents numerous practical benefits. This approach provides crucial metabolic diversity and resilience, which are vital for successful application in real-world scenarios. Reports suggest about successfully degraded diesel oil and phenanthrene, resulting in a 60% decrease in isoprenoid levels and an overall reduction of approximately 75% in total hydrocarbons over a period of 42 days. This feat was accomplished through the use of a microbial blend consisting of carefully selected native strains. Generating the requisite biomass for inoculation in bio-augmentation is typically achieved within bioreactors, though transferring these cultures to the site often presents challenges. Microbial inoculants refer to uniform suspensions of cells that are prepared in optimal conditions. However, they frequently undergo stress when exposed to the intricacies of natural surroundings. In real-world circumstances, the introduced microbial population rapidly declines shortly after application owing to a variety of abiotic and biotic stressors. These stressors comprise fluctuations or extremes in temperature, moisture levels, pH, nutrient depletion, as well as conceivably dangerous levels of pollutants in the tainted soil. The utilization of carrier materials frequently provides physical reinforcement for biomass, while simultaneously augmenting the availability of nutrients, moisture, and aeration. This prolongation of microbial viability is attributed to the enhanced circumstances afforded. Microbial cell encapsulation or immobilization, in particular, can yield heightened survival rates by safeguarding cells against stressful environmental factors. This shielding effect typically results in accelerated and more efficient biodegradation when contrasted with freely suspended cells. A multitude of materials, namely agar, acrylate copolymers, alginate, agarose, kappa-carrageenan,, polyurethane, and polyvinyl alcohol gel, gelatin, gellan gum, have undergone rigorous scrutiny and experimentation for the purpose of cell encapsulation or immobilization (Cassidy et al. 1996).

4.6.2 Bio Stimulation

Studies have demonstrated that the deterioration of petroleum hydrocarbons by indigenous microbial communities can be enhanced with the presence of vital nutrients within the contaminated surroundings. Spills of petroleum hydrocarbons provide a significant carbon source for native microorganisms, yet nitrogen and phosphorus are often scarce in many environments. Consequently, bio-stimulation expedites the decontamination process by introducing one or more of these limiting nutrients. This supplementation improves the degradation capacity of the existing microbial population (Nikolopoulou and Kalogerakis 2009). Environmental circumstances can also be fine-tuned, including the regulation of pH and moisture levels, to achieve optimal conditions for microbial degradation. Numerous scientists have investigated the enhancement of nutrient levels by introducing nitrogen and phosphorus. For instance, few investigations have achieved a substantial enhancement of petroleum hydrocarbon biodegradation, up to 96%, by introducing bio-solids (nutrient-rich organic matter derived from domestic sewage treatment) and inorganic fertilizers (rich in nitrogen and phosphorus) into diesel-contaminated soils (Sarkar et al. 2005). In addition to nutrients, numerous other factors exert substantial influence on the degradation rate of petroleum hydrocarbons (PHCs) in natural environments. For instance, in the aftermath of the Exxon Valdez oil spill, it was noted that the extent of physical blending, initial treatments, and the presence of alternative carbon sources had an effect on the mineralization abilities of microorganisms (Sugai et al. 1997). Horel and Schiewer conducted a research study that aimed to determine the impact of varying temperature and moisture levels on the bio-stimulation of Syntroleum through the use of fertilizer as a stimulating agent. The researchers observed that the bioremediation process was initiated earlier at higher temperatures (20 °C) in comparison to lower temperatures (6 °C) for shorter incubation periods (4–6 weeks). However, over extended incubation periods (12–17 weeks), microbes adapted to lower temperatures and managed to degrade around 50% of Syntroleum. In all scenarios, the presence of nutrients was found to be crucial for the microbial breakdown of substances in the soil. After a 17 week period, the degradation rates were approximately three times greater at a temperature of 20 °C and eight times higher at 6 °C compared to soils lacking essential nutrients. Conversely, alterations in moisture levels (which ranged from 2 to 12% by weight) and the routine process of mixing the soil to improve its porosity did not have a significant impact on the degradation process. Evidently, the amalgamation of bio-augmentation, bio-stimulation, and the incorporation of bio-surfactants, tailored to the specifics of the polluted area, shows potential for expediting remediation processes. Nonetheless, any such deliberate intervention necessitates subsequent assessments of ecotoxicity and site quality. These evaluations are essential to ensure the restoration of the contaminated site's original biological vitality and overall integrity (Liu et al. 2009).

4.6.3 Constructed Wetlands

Constructed wetlands possess the capacity to establish the necessary conditions for microbial dehalogenation. These wetlands can be categorized into two types: surface flow (SF) and subsurface flow (SSF). In surface flow wetlands, water predominantly moves above the ground surface, flowing as shallow sheets over the dense growth of emergent wetland vegetation (Knight et al. 1999). The design elements encompass an inlet mechanism, the wetland basin, the wetland flora, and an outlet mechanism. The size, shape, and quantity of basins are determined by the practical reaction kinetics needed to meet regulatory discharge standards. The number of basins is influenced by flow rates, land limitations, and the need for operational backup. Wetland plants facilitate mineral cycling and offer surfaces for microbial populations to adhere to. Plant selection considers their resilience under expected conditions, their capacity to absorb contaminants (if applicable), cost-effectiveness, availability, value as wildlife habitat and nourishment sources. The outlet mechanism gathers surface water and directs it to another wetland cell or to the final discharge destination like a stream, river, lake, or groundwater replenishment site. The size and types of plant life within a natural wetland are defined by the quantity and quality of water present. The hydrological conditions in turn impact the soil and nutrient availability, influencing the characteristics of microbial communities. Natural wetlands receive water through streamflow, runoff, groundwater discharge, and precipitation (Kadlec and Reddy 2001). In constructed wetlands, water intake is managed by an input mechanism, while water exits through streamflow, groundwater replenishment, and evapotranspiration. For constructed wetlands, an output mechanism and evapotranspiration facilitate water discharge. Natural wetlands exhibit a variable capacity for water storage, ranging from substantial to minimal amounts. Alterations in water depth during different seasons can affect the types of species present and the microbial communities within the sediment. The underlying terrain can lead to diverse zones within the wetland, with some areas being susceptible to drying out (Caswell et al. 1992).

4.6.4 Bio-pile Systems

It constitutes a blend of composting and land farming techniques. The fundamental bio pile setup consists of a treatment bed, aeration arrangement, nutrient-infusion system, and leachate collection mechanism. To ensure optimal degradation, it is imperative to regulate various factors such as moisture, temperature, nutrients, oxygen levels, and pH. The subterranean irrigation system is concealed within the soil, and a vacuum mechanism is employed to introduce air and nutrients. In order to mitigate runoff, a plastic covering is employed over the soil surface, which concurrently curbs vaporization and promotes solar heating. The bio-pile treatment process generally spans 2–3 months for completion (Benyahia and Embaby 2016). Bio piles, which are also known as biomounds, bioheaps, biocells, and static pile composts, combine aspects of both land-farming and composting methods. This method is becoming more commonly used for the remediation of excavated soils contaminated with petroleum compounds, employing biodegradation processes. What distinguishes it from land-farming is its ability to carefully control optimal levels of moisture, temperature, pH, aeration, and nutrients, all of which promote the thriving presence and activity of microorganisms, thus accelerating rapid biodegradation. In bio-piling, contaminated soils are assembled into piles or heaps, and then microbiological processes are invigorated through aeration, subsequent water and nutrient supplementation, while also regulating heat and pH conditions (Hennebel et al. 2009).

In contrast to land-farming or composting methods, bio-piles potentially provide a more efficient strategy for removing pollutants due to better mass transfer capabilities for air, nutrients, and water. Bio-piles implement structural components such as sand, straw, sawdust, woodchips, and desiccated manure in order to optimize the efficiency of mass transfer. Bio-piles can be designed as enclosed systems to manage vapour emissions and can be engineered to effectively address a range of petrochemical compounds and environmental conditions. However, despite being an environmentally friendly and cost-effective technology, bio-piling does come with certain limitations. It might not be suitable for highly concentrated contaminant scenarios (Li et al. 2004). Coulon et al. carried out an empirical investigation in the former dockyard of Scotland, aimed at comparing the efficacy of the bio-piling and windrow turning techniques for the remediation of the soil contaminated with bunker C fuel. The outcomes of the study indicated that windrowing was more effective in treating bunker fuel soils because of their friable nature, whereas bio-piling was found to be better suited for coarser soil textures (Coulon et al. 2010).

4.7 Case Studies and Field Applications

Reaction to Marine Oil Spills

In responding to marine oil spills, it's important to note that crude oil exhibits a strong aversion to water, resulting in very limited solubility despite being composed of numerous hydrocarbons, nonhydrocarbon components, and metals, each with their unique solubility in aqueous environments. When crude oil encounters water, it stays afloat on the surface of the water because it has a lower density than water. Surface Tension (ST), in conjunction with viscosity, serves as an indicator of how quickly and extensively the oil spreads across the water's surface and, once dispersed, within the underwater environment. A decrease in the interfacial tension that exists between oil and water has the consequence of promoting greater diffusion of the oil over the surface of the water. Generally, when oil is dispersed, it takes the form of small droplets that are neutrally buoyant and possess a larger surface area compared to their volume. These droplets typically range in size from 1 to 70 μm. This characteristic enhances the accessibility of the oil to hydrocarbon-degrading bacteria, facilitating its biodegradation. The comprehensive documentation of the natural course of biodegradation of crude oil, which occurs via microbial biological oxidation, within marine ecosystems has been established (Kostka et al. 2011). The primary emphasis in the realm of marine bioremediation has been on the utilization of fertilizers as well as the introduction of seed cultures that comprise highly potent microorganisms capable of effectively degrading oil. However, this approach has generated varying degrees of success (Timmis et al. 2010). The objective of incorporating surfactants, whether naturally occurring or artificially synthesized, is to facilitate the dispersion/emulsification of oil and, consequently, accelerate the biodegradation process. It has been demonstrated that biosurfactants are efficacious in dispersing unrefined petroleum and augmenting the biodegradation process, but solely under controlled laboratory settings, owing to the logistical, financial, and regulatory hindrances impeding extensive field trials. In a meticulously executed laboratory investigation, a synergistic assemblage of bacteria comprising *Ochrobactrum* and *Brevibacillus* strains, both with and without the incorporation of rhamnolipids, was introduced into a mixture of crude oil and one liter of water, mimicking the bioremediation process for marine oil spills. The findings divulged a fascinating observation: when the bacterial consortium worked in isolation, its efficacy in removing oil was 6% lower in comparison to its performance when supplemented with bio-surfactants. Significantly, the presence of rhamnolipids demonstrated a remarkable ability to augment the biodegradation of longer-chain alkanes (>nC_{15}), whereas interestingly, it had the opposite effect on shorter-chain alkanes (nC_{13}–C_{15}). In such instances, rhamnolipids reduced solubility and heightened stability. Nevertheless, the overall efficiency of eliminating n-alkanes through the combined action of the bacterial consortium and rhamnolipids surpassed that of the control. This intriguing trend also extended to the elimination of PAHs and biomarkers (Chen et al. 2013). The results presented in this study demonstrate a significant degree of consistency with a prior investigation that utilized rhamnolipids in conjunction with a preadapted bacterial consortium. Additionally, it was observed that the effects of rhamnolipids were greatly enhanced when nutrients were introduced into the experimental treatments. The reported specific degradation rates for nC_{15}, nC_{20}, and nC_{25} hydrocarbons were notably higher than the control, with values of 23, 20, and 10 times greater, respectively. This outcome is certainly impressive. Furthermore, it was found that rhamnolipids exhibit the ability to stimulate the proliferation of hydrocarbon-degrading microorganisms within a short period of just 5 days. This led to a notable utilization of 50% of the crude oil's saturated fraction. Moreover, it was observed that this surfactant has a significant impact on the degradation of low-molecular-weight PAHs, as well as on the biomarkers pristane and phytane (Nikolopoulou et al. 2013). The source of nutrients, whether they are organic, lipophilic, or water-soluble, when combined with rhamnolipids in treatment scenarios, holds the potential to further augment crude oil degradation in both marine water and sediment environments (McKew et al. 2007). In the domain of oil spill response, bio-emulsifiers have emerged as a promising tool.

Specifically, EPS2003, an exopolysaccharide bio-emulsifier produced by *Acinetobacter calcoaceticus*, has garnered attention. EPS2003 has demonstrated its effectiveness in enhancing the biodegradation of crude oil in natural seawater microcosms. The introduction of EPS2003 into the microcosms has yielded a twofold increase in the overall oil biodegradation rate, as well as an increase in the population of hydrocarbon-degrading bacteria such as *Alcanivorax*, *Marinobacter*, *Oceanospirillum*, and *Pseudomonas*, when compared to microcosms without EPS (Cappello et al. 2012).

There is a noticeable scarcity of field studies assessing the practical effectiveness of bio-surfactants in treating oil spills. This scarcity is primarily due to factors such as the high market costs of these bio-surfactants and logistical challenges associated with field research. During a laboratory experiment, the scientific community endeavored to conduct a comparative analysis between the synthetic chemical dispersant Ultrasperse II and surfactin, a naturally occurring substance derived from the *Bacillus* sp. strain H_2O-1. The study's results indicate that surfactin had a more significant impact on the hydro-carbonoclastic bacterial population compared to the synthetic dispersant. However, there was no observable difference in oil biodegradation between the two treatments. A recent study conducted a comparison between rhamnolipid, trehalolipid, and sophorolipid biosurfactants and three commercial dispersants in controlled environments contaminated with oil in marine coastal water. The results showed that while the biosurfactants yielded unique microbial responses, they were equally effective in promoting alkane biodegradation as the microcosms treated with dispersants. In a prior study, researchers compared rhamnolipid produced by Pseudomonas aeruginosa with the widely used synthetic chemical dispersant Finasol OSR52. The present investigation concentrated on evaluating the impact of crude oil exposure on a naturally occurring bacterial population obtained from the Faroe-Shetland Channel, a subarctic area situated in the Northeast Atlantic (Nikolova et al. 2021). Significantly, the application of rhamnolipid led to a higher variety of bacteria capable of degrading oil in comparison to the synthetic dispersant. However, it was noted that the overall degradation of crude oil, with the exception of its aromatic portion, showed a more noticeable improvement when the synthetic dispersant was utilized with the oil.

Zhang et al. tackled the long-standing problem of paraffin build up in the oil industry. While mechanical and chemical methods have their issues, they explored a microbiological approach involving fungi, which hasn't been widely studied compared to bacteria. In a laboratory experiment, six different strains of Aspergillus were tested, and it was observed that their enzyme preparations effectively increased the solubility of paraffin wax in n-hexane and broke down its constituents. During this process, gases such as CO_2 and H_2, along with organic acids like oxalate and propionate, were generated. Additionally, the structure of the wax crystals underwent a transformation, becoming rougher and more porous. This research underscores the potential utility of employing extracellular enzymes derived from Aspergillus spp. to effectively address issues associated with paraffin wax in oil wells. The enzyme preparations obtained from these six fungal cultures exhibited promising capabilities in degrading paraffin wax, making them a viable choice for mitigating wax-related challenges in the industry (Zhang et al. 2015).

Microbial Enhanced Oil Recovery (MEOR)

Microbial Enhanced Oil Recovery (MEOR) involves introducing microorganisms or their byproducts into existing oil reservoirs to recover previously unrecovered crude oil. This method relies on favorable reservoir conditions for microorganism growth and their metabolic products to aid in oil mobilization. Lipopeptides, a type of biosurfactant, exhibit great potential for use in Microbial Enhanced Oil Recovery (MEOR) applications. Their ability to effectively lower Interfacial Tension (IFT) to levels below 0.1 mN/m has been demonstrated through successful lab and in-situ MEOR investigations, and they are predominantly generated by Bacillus spp. strains. Surfactins, a type of lipopeptide, exhibit efficacy across various environmental conditions and have proven effective in oil recovery from sand. B. subtilis-produced surfactin emulsifies diesel efficiently, even at high temperatures, and can recover over 60% of trapped oil in sand cores. Surfactin's potential extends to altering CO_2 wettability in subsurface rock formations, making it suitable for carbon capture and storage (Park et al. 2017). Lichenysin, another biosurfactant, reduces IFT to levels below 10 mN/m, even in harsh conditions like high temperature, pH range of 6–10, up to 10% NaCl, and calcium concentrations up to 340 mg/L. In the realm of core flooding experiments, the utilization of partially purified lichenysin resulted in a notable retrieval of up to 40% of residual oil from sandstone cores, surpassing the efficacy of chemical surfactants. Combining biosurfactants with chemical surfactants enhances performance. Rhamnolipids reduce surfactant adsorption to sandstone, improving oil recovery quality by 7%. Macromolecular biopolymers, such as emulsan, have the ability to eliminate a significant proportion, up to 98%, of preadsorbed crude oil from limestone cores, even when present in low concentrations of 0.5 mg/mL. In addition, biopolymers sourced from Rhizobium viscosum CECT908 are known to outperform xanthan gum in terms of heavy oil recovery (Nikolova and Gutierrez 2023).

Soil Bioremediation

Soil bioremediation is a solution to the prevalent issue of soil contamination due to hydrocarbons and heavy metals resulting from various human activities such as drilling, pipeline leaks, storage tank incidents, and transportation. These contaminants, particularly when attached to soil particles, are highly resistant to elimination (Youssef et al. 2009). While conventional methods such as eradication, incineration, soil washing, and solvent extraction have exhibited effectiveness in previous applications, they have been revealed to have unfavorable impacts on soil structure and local biodiversity, and are often linked with exorbitant costs. In contrast, this alternative approach is predicated on leveraging indigenous soil microorganisms to facilitate the transformation of petroleum hydrocarbons into microbial biomass, carbon dioxide, and water. Numerous factors impinge upon the degree and pace of hydrocarbon degradation in soil, including moisture content, aeration, pH, temperature, soil composition (including nutrient richness or aging and bioavailability), as well as the concentration, molecular structure, and bioavailability of the hydrocarbon contaminants. Properly optimizing these environmental factors is crucial for successful bioremediation. Soil bioremediation can be conducted either on-site (in situ) or by excavating contaminated soil for treatment elsewhere (ex situ). In in situ bioremediation, typically, the top 30 cm layer of soil is treated with fertilizers to stimulate the activity of native soil microorganisms responsible for breaking down hydrocarbons (Huesemann et al. 2004). While prioritizing this methodology, it is imperative to take into account the plausible hazard of tainting the subsurface water-bearing strata with disintegrated hydrocarbons. Occasionally, incompletely refined bio-surfactants have been implemented in situ bioremediation to augment the hydrocarbon solubility and bio-accessibility by expanding their surface area, inclusive of other hydrophobic toxicants. A significant trial was carried out on LaTouche Island situated in Alaska, which showcased the potential of PES-51, a surfactant that is biologically derived, to effectively remove nearly one-third of semi-volatile hydrocarbons derived from petroleum found in subterranean beach materials. A study carried out in Argentina has revealed that surfactin, originating from the B. subtilis strain O9, has significantly enhanced the removal of crude oil from sandy loam soil over a 300 day period (Cubitto et al. 2004). Additionally, in another instance, the incorporation of rhamnolipid that hails from Pseudomonas aeruginosa strain SSC2 into soil sludge that is contaminated with crude oil has resulted in a remarkable 98% degradation within a mere 4 weeks, in contrast to the non-rhamnolipid control treatment that attained 67% degradation. This effect was further augmented by the supplementation of nutrients to the treatment (Cameotra and Singh 2008). Moreover, a soil bioremediation experiment conducted in the vicinity of an oil production plant in Pakistan has demonstrated that a blend of a unique bacterial consortium, rhamnolipids, and nutrients has resulted in elevated rates of crude oil degradation, reaching as high as 77%. The efficaciousness of Mannosylerythritol lipids (MELs) that were produced by *Candida antarctica* SY16 in the degradation of crude oil in soil was scrutinized by Baek et al. (2007). A consortium of microorganisms capable of degrading hydrocarbons was introduced into soil contaminated with crude oil. Diverse combinations of rhamnolipids and nutrients were supplemented. The soil that received this combination exhibited the highest degree of crude oil degradation, achieving a percentage of 77.6%. Furthermore, the incorporation of these supplementary substances ameliorated the survival and proliferation of microorganisms in the soil, as well as the expression of the alkane hydroxylase gene, alkB, which is implicated in the breakdown of oil. A robust positive correlation (with a correlation coefficient of 0.94) between gene expression and the depletion of crude oil implies that the expression of catabolic genes is pivotal for the breakdown of hydrocarbons. Taken together, these findings evince that the addition of rhamnolipids and nutrients heightens microbial colonization and metabolic activity, consequently enhancing the remediation of soil contaminated with crude oil (Tahseen et al. 2016).

Nur Aina Adlan and colleagues have demonstrated in their research that *Geobacillus kaustophilus* N3A7, *Geobacillus kaustophilus* DFY1, *Geobacillus jurassicus* MK7, *Geobacillus stearothermophilus* AD11, and *Geobacillus stearothermophilus* AD24 are capable of efficiently degrading crude oil with a biodegradation efficiency of over 70%. These strains are believed to produce degradative enzymes that facilitate the uptake and degradation of hydrocarbons by bacterial cells. The enzymatic degradation of long-chain n-alkane, which involves cleaving large aliphatic chains or internal linkages to produce short-chain n-alkane, as well as the possibility of generating short-chain n-alkanes through decarbonylation, suggest that these strains may be useful for reducing the pour point of crude oil and accelerating the biodegradation process (Adlan et al. 2020b).

Studies conducted by Tian and colleagues collected six strains from heavily oil-contaminated soil in Bohai Bay, China, which exhibited highly efficient hydrocarbon degradation capabilities. These strains were identified as belonging to the Bacillus sp. and Pseudomonas sp. The combination of these strains, referred to as YJ01, demonstrated robust survival and effective petroleum degradation (reaching 80.64%), highlighting a synergistic effect among the isolates. Additionally, YJ01 displayed the ability to degrade high concentrations of crude oil (15 g/L) with a rapid degradation rate (half-life of 5.48 days). These strains not only broke down short-chain alkanes but also tackled more complex long-chain alkanes, showcasing their versatility. Furthermore, they exhibited enhanced paraffin degradation activity. Consequently, these isolated strains hold promise for applications in remediating crude oil-contaminated soil and treating oily wastewater.

4.8 Evaluating the Sustainability and Efficiency of Paraffin-Degrading Bioremediation

Adlan et al. conducted a study on the biodegradation of crude oil paraffin wax using eleven bacterial strains that were isolated from seawater and oil-contaminated soil samples. The results showed that strains N3A7, MK7, DFY1, AD11, and AD24 were able to achieve more than 70% biodegradation efficiency of crude oil in just 3 days. During the initial 2 days of incubation, most of the strains were able to completely degrade C_{37}–C_{40} and increase the ratio of C_{14}–C_{18}. Furthermore, the degradation of crude oil led to changes in the pH of the medium. Strain AD24 exhibited the highest activities of alkane monooxygenase, while strains N3A7, MK7, DFY1, AD11, and AD24 showed higher overall alcohol dehydrogenase, lipase, and esterase activities. These findings suggest that these bacterial strains have the potential to be used in controlling wax deposition and improving the quality of oil. It was found that strains N3A7, MK7, DFY1, AD11, and AD24 had higher biodegradation efficiency of long-chain n-alkanes and increased short-chain alkanes, particularly on days 1 and 2. The efficient degradation of crude oil by these strains could be a viable solution for controlling wax deposition and improving oil flow in pipelines. The study found that certain bacterial strains, including *Geobacillus kaustophilus* N3A7, *Geobacillus kaustophilus* DFY1, *Geobacillus jurassicus* MK7, *Geobacillus stearothermophilus* AD11, and *Geobacillus stearothermophilus* AD24, showed high efficiency in degrading crude oil, with more than 70% biodegradation efficiency. These strains produced degradative enzymes, such as alkane monooxygenase, alcohol dehydrogenase, lipase, and esterase, which played a role in the degradation process. The increase in alkane monooxygenase activity corresponded to the increase in crude oil degradation efficiency. The enzymatic degradation of long-chain n-alkanes into short-chain n-alkanes by these strains could potentially reduce the pour point of crude oil and accelerate the biodegradation process. These findings suggest the potential application of these bacterial strains in controlling wax deposition and improving the quality of oil.

Zhang et al. (2015) conducted an investigation on crude enzyme preparations derived from six fungal cultures, namely *Aspergillus fumigatus*, *A. flavus*, and *A. terreus*. The said preparations exhibited commendable activities in the degradation of paraffin wax. The enzyme preparations demonstrated a notable capability to effectively degrade paraffin wax, resulting in a substantial increase in its solubility in n-hexane. This degradation process was accompanied by the generation of gaseous byproducts such as CO_2 and H_2, in addition to the emergence of organic acids including oxalate and propionate. As a consequence of enzymatic degradation, the morphology of the wax crystals underwent a marked transformation, leading to the development of a rough surface and a loose structure. The investigation emphasizes the capability and practicality of implementing fungal enzyme formulations, specifically those sourced from Aspergillus spp., for the purpose of eliminating wax. Future undertakings ought to concentrate on evaluating the suitability of raw Aspergillus enzyme formulations in oil wells, particularly those with greater paraffin breakdown proficiency.

Patel et al.'s (Patel and Bhaskaran 2018) research found that Pseudomonas nitroreducens, which was obtained from oil wells in Gujarat, India, demonstrated a substantial capacity for degrading paraffin wax, particularly in its major components of heneicosane, pentacosane, and triacontane. The isolate, identified as Pseudomonas nitroreducens, further exhibited efficient degradation of paraffinic crude oil, with a degradation rate of 70% within a 10 day period. Gas chromatographical analysis confirmed the ability of Pseudomonas nitroreducens to degrade paraffin wax, with complete utilization of heneicosane, significant utilization of pentacosane, and triacontane after 8 days of incubation. The research suggests that Pseudomonas nitroreducens holds great promise as a tool for the remediation of paraffin wax in oil wells, providing an ecologically sound and cost-effective approach.

4.9 Challenges and Limitations

Petroleum hydrocarbons pose a considerable threat because of their substantial harm to both the environment and human health. Bioremediation, which utilizes bacteria capable of degrading petroleum hydrocarbons, is widely recognized as an environmentally sustainable and efficient technology. Numerous bacterial species with the capability to degrade petroleum hydrocarbons have been identified and utilized in bioremediation efforts. However, practical applications have encountered several challenges that impede the biodegradation process (Nikolova and Gutierrez 2021). To address these challenges, various strategies have been explored and implemented, including the regulation of environmental factors and the optimization of microbial inoculants. Despite the progress reviewed here, there remains a need for further research before bioremediation can be successfully applied to restore petroleum oil-contaminated environments. The following conclusions can be drawn:

1. Enhancing our knowledge of the theoretical aspects related to how bacteria and petroleum hydrocarbons interact at the interface is crucial for overcoming obstacles that impede the microbial absorption of these hydrocarbons.
2. The creation of novel surfactants that are compatible with biological systems is of utmost importance for improving the interaction between bacteria and petroleum hydrocarbons.
3. Exploration of untapped reservoirs of petroleum hydrocarbon-degrading bacteria through cutting-edge biotechnological approaches, including high-throughput screening techniques, has the potential to broaden and enhance the range of functional bacterial resources available.
4. Further optimization of artificial microbial consortia, potentially through metagenome enrichment approaches, is needed to create more effective and preferred consortia.
5. Exploring innovative functional genes responsible for regulating hydrocarbon degradation pathways can offer new perspectives into the molecular mechanisms that govern microbial remediation.
6. Leveraging synthetic biology technology to engineer bacteria with enhanced abilities for petroleum hydrocarbon degradation represents another avenue for exploration.

These measures are essential for advancing the effectiveness of bioremediation in restoring petroleum-contaminated environments (Xu et al. 2018).

4.10 Future Prospects and Conclusion

In recent times, substantial attention has been directed towards bio-surfactants and bio-emulsifiers, owing to the escalating demand for natural ingredients by consumers and the industries' quest for functional constituents from sustainable sources. The extensive marine ecosystem, recognized for its opulent biosphere and diverse microbial community, presents a significant opportunity for the exploration of innovative biomolecules with diverse industrial applications. In the Oil and Gas sector, there is a marked interest in exploiting biosurfactants for Microbial Enhanced Oil Recovery (MEOR) and developing dispersants for oil spill remediation, although these applications have not yet attained widespread industrial adoption. Despite significant progress in comprehending the harmful impact of dispersants on marine life since their initial examination in the 1960s, persistent debates and discussions continue to exist. Numerous inquiries have been carried out in order to scrutinize the consequences of dispersants. However, the results have been inconsistent, in spite of employing the identical type of dispersant for different research endeavours. This realm calls for increased emphasis to establish credibility among both the scientific society and the general populace regarding their application. In the end, the benefits of dispersants should surpass any potential harm, whether immediate or prolonged, to the marine ecosystem and human well-being. Recent advancements in metagenomics techniques have presented prospects for exhaustively delineating autochthonous microbial populations in oil reservoirs, encompassing their taxonomic classification and functional attributes, to an unparalleled level of precision. The utilization of metagenomics in comprehending the microbiology of oil reservoirs is predicted to unveil hitherto undiscovered microbial variants, thereby providing opportunities to exploit their latent potential for Microbially Enhanced Oil Recovery (MEOR). The confluence of biotechnological methodologies, specifically recombinant DNA techniques, holds great potential for augmenting Microbial Enhanced Oil Recovery (MEOR). Contemporary strategies for Microbial Enhanced Oil Recovery (MEOR) are considerably reliant on genetic engineering techniques, alongside the in situ and ex situ cultivation of a range of organisms, the biosynthesis of cellular products, and the kinetics of microbial enzymes. It is widely acknowledged that the enhancement of Microbial Enhanced Oil Recovery (MEOR) requires thorough investigation and isolation of unexplored microbial strains, particularly those capable of thriving in anaerobic environments and exhibiting accelerated production of biomass or surface-active biopolymers.

With the development of genetically engineered bacterial strains, greater efficiencies could be attained, especially in terms of ex situ biosurfactant and/or biopolymer production, or the more effective degradation of high-molecular-weight hydrocarbons found in heavy crude oils, thereby rendering these oils more accessible in reservoirs. However, the release of genetically modified organisms into the environment is restricted by regulatory constraints and safety considerations. In view of the pressing issue of global climate change that poses a threat to life on Earth due to unrestrained greenhouse gas emissions, and considering the limited progress made in MEOR technology in recent decades, there is an urgent need for substantial investment in the research and development of technologies that focus on extracting energy from non-fossil fuel sources.

Acknowledgements The authors acknowledge their department and Institute for providing essential facilities and support.

Author Contributions SP, SS and SD contributed equally to the conception of the study, manuscript writing, and editing. US contributed significantly to the manuscript writing, and illustration preparation. RP and DDM wrote the manuscript with constructive discussions.

Consent for Publication All authors have consented to the manuscript for publication.

References

Adlan NA, Sabri S, Masomian M et al (2020a) Microbial biodegradation of paraffin wax in Malaysian crude oil mediated by degradative enzymes. Front Microbiol 11:565608. https://doi.org/10.3389/fmicb.2020.565608

Adlan NA, Sabri S, Masomian M et al (2020b) Microbial biodegradation of paraffin wax in Malaysian crude oil mediated by degradative enzymes. Front Microbiol 11

Agrawal SK, Hoch LE (2016) Saturating wax coating composition and associated methods of use

Agrawal SK, Hoch LE (2019) Saturating wax coating composition and associated methods of use

Allison R (1998) A note on paraffin waxes, their crystals, and microtoming. Microsc Today 6:18–19. https://doi.org/10.1017/S1551929500067250

Analysis of Pseudomonas putida alkane-degradation gene clusters and flanking insertion sequences: evolution and regulation of the alk genes. The EMBL accession numbers for the sequences reported in this paper are AJ245436 [P. putida (oleovorans) GPo1 alk gene clusters and flanking DNA], AJ233397 (P. putida P1 alk gene clusters and flanking DNA), AJ249793 (P. putida P1 nahKJ genes), AJ249825 [P. putida (oleovorans) GPo1 16S RNA gene] and AJ271219 (P. putida P1 16S RNA gene). Microbiology Society. https://doi.org/10.1099/00221287-147-6-1621. Accessed 4 Sep 2023

Andre D, Contremoulins S, Jolivet Y (2004) Paraffin wax composition useful for making candles comprises soft wax and an elastomeric block copolymer and/or microcrystalline wax

Baek K-H, Yoon B-D, Kim B-H et al (2007) Monitoring of microbial diversity and activity during bioremediation of crude oil-contaminated soil with different treatments. J Microbiol Biotechnol 17:67–73

Beilen JBV, Li Z, Duetz WA et al (2003) Diversity of alkane hydroxylase systems in the environment. Oil Gas Sci Technol 58:427–440. https://doi.org/10.2516/ogst:2003026

Benyahia F, Embaby AS (2016) Bioremediation of crude oil contaminated desert soil: effect of biostimulation, bioaugmentation and bioavailability in biopile treatment systems. Int J Environ Res Public Health 13:219

Biodegradation and bioremediation of hydrocarbons in extreme environments. SpringerLink. https://doi.org/10.1007/s002530100701. Accessed 4 Sep 2023

Bognolo G (1999) Biosurfactants as emulsifying agents for hydrocarbons. Colloids Surf Physicochem Eng Asp 152:41–52

Cameotra SS, Singh P (2008) Bioremediation of oil sludge using crude biosurfactants. Int Biodeterior Biodegrad 62:274–280. https://doi.org/10.1016/j.ibiod.2007.11.009

Cappello S, Genovese M, Della Torre C et al (2012) Effect of bioemulsificant exopolysaccharide (EPS2003) on microbial community dynamics during assays of oil spill bioremediation: a microcosm study. Mar Pollut Bull 64:2820–2828

Cassidy MB, Lee H, Trevors JT (1996) Environmental applications of immobilized microbial cells: a review. J Ind Microbiol Biotechnol 16:79–101

Caswell PC, Gelb D, Marinello SA et al (1992) A component performance evaluation of constructed surface flow and wetlands cells for produced water treatment in the pitchfork field, Wyoming. In: SPE annual technical conference and exhibition? SPE, p 24808

Chen Q, Bao M, Fan X et al (2013) Rhamnolipids enhance marine oil spill bioremediation in laboratory system. Mar Pollut Bull 71:269–275

Coulon F, Al Awadi M, Cowie W et al (2010) When is a soil remediated? Comparison of biopiled and windrowed soils contaminated with bunker-fuel in a full-scale trial. Environ Pollut 158:3032–3040

Cubitto MA, Morán AC, Commendatore M et al (2004) Effects of Bacillus subtilis O9 biosurfactant on the bioremediation of crude oil-polluted soils. Biodegradation 15:281–287. https://doi.org/10.1023/b:biod.0000042186.58956.8f

Degradation of polycyclic aromatic hydrocarbons and long chain alkanes at 6070 °C by Thermus and Bacillus spp. SpringerLink. https://doi.org/10.1023/A:1027357615649. Accessed 4 Sep 2023

DeLong EF (2006) Archaeal mysteries of the deep revealed. Proc Natl Acad Sci 103:6417–6418

El Fantroussi S, Agathos SN (2005) Is bioaugmentation a feasible strategy for pollutant removal and site remediation? Curr Opin Microbiol 8:268–275

Etoumi A (2007) Microbial treatment of waxy crude oils for mitigation of wax precipitation. J Pet Sci Eng 55:111–121

Farsani RY, Raeiszadeh F, Jahangiri M, Afrand M (2020) Melting characteristics of paraffin wax in a rectangular cavity under steady rotations. J Taiwan Inst Chem Eng 113:135–141

Feng GL, Liu YZ, Jiao WZ et al (2011) Preparation and property of paraffin wax emulsion. China Surfactant Deterg Cosmet 41:415–418

Hao D-H, Lin J-Q, Song X et al (2008) Isolation, identification, and performance studies of a novel paraffin-degrading bacterium of Gordonia amicalis LH3. Biotechnol Bioprocess Eng 13:61–68

Hennebel T, Simoen H, De Windt W et al (2009) Biocatalytic dechlorination of trichloroethylene with bio-palladium in a pilot-scale membrane reactor. Biotechnol Bioeng 102:995–1002

Hsu CS, Robinson PR (2019) Classification and characterization. In: Hsu CS, Robinson PR (eds) Petroleum science and technology. Springer International Publishing, Cham, pp 57–64

Huesemann MH (2004) Biodegradation and bioremediation of petroleum pollutants in soil. In: Singh A, Ward OP (eds) Applied bioremediation and phytoremediation. Springer, Berlin, Heidelberg, pp 13–34

Ingham CJ, Sprenkels A, Bomer J et al (2007) The micro-Petri dish, a million-well growth chip for the culture and high-throughput screening of microorganisms. Proc Natl Acad Sci 104:18217–18222

Kadlec RH, Reddy KR (2001) Temperature effects in treatment wetlands. Water Environ Res 73:543–557

Kauppi S, Sinkkonen A, Romantschuk M (2011) Enhancing bioremediation of diesel-fuel-contaminated soil in a boreal climate: comparison of biostimulation and bioaugmentation. Int Biodeterior Biodegrad 65:359–368

Knight RL, Kadlec RH, Ohlendorf HM (1999) The use of treatment wetlands for petroleum industry effluents. Environ Sci Technol 33:973–980

Koma D, Hasumi F, Yamamoto E et al (2001) Biodegradation of long-chain n-paraffins from waste oil of car engine by Acinetobacter sp. J Biosci Bioeng 91:94–96. https://doi.org/10.1016/S1389-1723(01)80120-1

Kostka JE, Prakash O, Overholt WA et al (2011) Hydrocarbon-degrading bacteria and the bacterial community response in Gulf of Mexico beach sands impacted by the Deepwater Horizon oil spill. Appl Environ Microbiol 77:7962–7974

Krasavtsev BE, Aleksandrova EA, Khadisova ZhT, Aleksandrov BL (2017) Thermal properties of wax compositions. Chem Technol Fuels Oils 53:360–368. https://doi.org/10.1007/s10553-017-0812-8

Lee K, Levy EM (1989) Biodegradation of petroleum in the marine environment and its enhancement. Aquatic toxicology and water quality management. Wiley, New York, pp 217–243

Leys NM, Bastiaens L, Verstraete W, Springael D (2005) Influence of the carbon/nitrogen/phosphorus ratio on polycyclic aromatic hydrocarbon degradation by Mycobacterium and Sphingomonas in soil. Appl Microbiol Biotechnol 66:726–736

Li L, Cunningham CJ, Pas V et al (2004) Field trial of a new aeration system for enhancing biodegradation in a biopile. Waste Manag 24:127–137

Liu YJ, Zhang AN, Wang XC (2009) Biodegradation of phenol by using free and immobilized cells of Acinetobacter sp. XA05 and Sphingomonas sp. FG03. Biochem Eng J 44:187–192

Lozupone CA, Knight R (2008) Species divergence and the measurement of microbial diversity. FEMS Microbiol Rev 32:557–578

Marchant R, Banat IM, Rahman TJ, Berzano M (2002) The frequency and characteristics of highly thermophilic bacteria in cool soil environments. Environ Microbiol 4:595–602

McKew BA, Coulon F, Yakimov MM et al (2007) Efficacy of intervention strategies for bioremediation of crude oil in marine systems and effects on indigenous hydrocarbonoclastic bacteria. Environ Microbiol 9:1562–1571

Megharaj M, Ramakrishnan B, Venkateswarlu K et al (2011) Bioremediation approaches for organic pollutants: a critical perspective. Environ Int 37:1362–1375
Misra S, Baruah S, Singh K (1995) Paraffin problems in crude oil production and transportation: a review. SPE Prod Facil 10:50–54
Mokhatab S, Towler B (2009) Wax prevention and remediation in subsea pipelines and flowlines. World Oil 230:55–58
New approaches for bioaugmentation as a remediation technology. https://doi.org/10.1080/10643380490452362. Accessed 22 Aug 2023
Nikolopoulou M, Kalogerakis N (2009) Biostimulation strategies for fresh and chronically polluted marine environments with petroleum hydrocarbons. J Chem Technol Biotechnol Int Res Process Environ Clean Technol 84:802–807
Nikolopoulou M, Pasadakis N, Kalogerakis N (2013) Evaluation of autochthonous bioaugmentation and biostimulation during microcosm-simulated oil spills. Mar Pollut Bull 72:165–173
Nikolova CN, Ijaz UZ, Magill C et al (2021) Response and oil degradation activities of a northeast Atlantic bacterial community to biogenic and synthetic surfactants. Microbiome 9:1–17
Nikolova C, Gutierrez T (2021) Biosurfactants and their applications in the oil and gas industry: current state of knowledge and future perspectives. Front Bioeng Biotechnol 9
Nikolova C, Gutierrez T (2023) Novel approaches in the use of biosurfactants in the oil industry and environmental remediation. In: Biosurfactants. Elsevier, pp 107–128
Oppenheimer CH, Hiebert FK (1989) Microbiological techniques for paraffin reduction in producing oil wells. Alpha Environmental Corp., Austin, TX (USA)
Park T, Joo H-W, Kim G-Y et al (2017) Biosurfactant as an enhancer of geologic carbon storage: microbial modification of interfacial tension and contact angle in carbon dioxide/water/quartz systems. Front Microbiol 8:1285
Patel DD, Bhaskaran L (2018) Study on paraffin wax degrading ability of Pseudomonas nitroreducens isolated from oil wells of Gujarat, India. Pet Sci Technol 36:583–590
Prevention and mitigation of paraffin deposition by biosurfactant-producing and paraffin-degrading Bacillus amyloliquefaciens strain 6-2c. ScienceDirect. https://www.sciencedirect.com/science/article/abs/pii/S1385894717317205. Accessed 4 Sep 2023
Raju MN, Scalvenzi L (2018) Petroleum degradation: promising biotechnological tools for bioremediation. In: Zoveidavianpoor M (ed) Recent insights in petroleum science and engineering. InTech
Response of the bacterial community in oil-contaminated marine water to the addition of chemical and biological dispersants. ScienceDirect. https://www.sciencedirect.com/science/article/abs/pii/S0301479716308234. Accessed 7 Sep 2023
Rodrigues CJ, de Carvalho CC (2019) Phenotypic adaptations help Rhodococcus erythropolis cells during the degradation of paraffin wax. Biotechnol J 14:1800598
Sarkar D, Ferguson M, Datta R, Birnbaum S (2005) Bioremediation of petroleum hydrocarbons in contaminated soils: comparison of biosolids addition, carbon supplementation, and monitored natural attenuation. Environ Pollut 136:187–195
Shahi A, Aydin S, Ince B, Ince O (2016) Evaluation of microbial population and functional genes during the bioremediation of petroleum-contaminated soil as an effective monitoring approach. Ecotoxicol Environ Saf 125:153–160
Siron R, Pelletier E, Brochu C (1995) Environmental factors influencing the biodegradation of petroleum hydrocarbons in cold seawater. Arch Environ Contam Toxicol 28:406–416
Sugai SF, Lindstrom JE, Braddock JF (1997) Environmental influences on the microbial degradation of Exxon Valdez oil on the shorelines of Prince William Sound, Alaska. Environ Sci Technol 31:1564–1572
Tahseen R, Afzal M, Iqbal S et al (2016) Rhamnolipids and nutrients boost remediation of crude oil-contaminated soil by enhancing bacterial colonization and metabolic activities. Int Biodeterior Biodegrad 115:192–198. https://doi.org/10.1016/j.ibiod.2016.08.010
Timmis KN, McGenity T, Van Der Meer JR, de Lorenzo V (2010) Handbook of hydrocarbon and lipid microbiology. Springer, Berlin

Wang W, Cui J, Wu S et al (2019) Isolation and optimization of compound bacteria for wax content and particle size. J Pet Sci Technol 9:53

Xiao M, Li W-H, Lu M et al (2012) Effect of microbial treatment on the prevention and removal of paraffin deposits on stainless steel surfaces. Bioresour Technol 124:227–232

Xu X, Liu W, Tian S et al (2018) Petroleum hydrocarbon-degrading bacteria for the remediation of oil pollution under aerobic conditions: a perspective analysis. Front Microbiol 9

Youssef N, Elshahed MS, McInerney MJ (2009) Chapter 6 microbial processes in oil fields: culprits, problems, and opportunities. In: Advances in applied microbiology. Academic Press, pp 141–251

Zhang J, Xue Q, Gao H, Wang P (2015) Biodegradation of paraffin wax by crude *Aspergillus* enzyme preparations for potential use in removing paraffin deposits: microbial removal of paraffin deposits. J Basic Microbiol 55:1326–1335. https://doi.org/10.1002/jobm.201500290

Zhang YL, Liu Z, Liu T (2016) Isolation and characterization of a novel paraffin wax-degrading bacterium, Pseudomonassp strain PW-1, from petroleum-contaminated sites. Genet Mol Res 15

Chapter 5
Selectively Isolated Biodegradable Bacteria for Deprivation of Organic Deposits in Oil and Gas Industries

Smita Lata, Swati Priya, Sagar Mondal, Indrani Barman, Sumira Malik, and Jutishna Bora

Abstract This chapter addresses organic deposit (or cumulose) secondary deposits that accumulate as a product primarily formed by the potential of the direct or indirect vital roles played by plant and animals remains and were preserved intact by a high-water table or by another circumstances preventing decomposition. The most common type of organic deposits includes coal, lignite, oil shale. In this we will mainly discuss about the organic deposit of oil and gas industries and their deprivation by biodegradable bacteria, oil and Gas are formed from organic materials mainly deposited as sediments on the seabed and broken-down and transformed over millions of years. Oil and gas extraction is a process that refers to the exploration and production of petroleum and natural gas from wells by method known as drilling. The formation of organic deposit is resulted by the precipitate of heavy hydrocarbons when temperature or pressure is reduced. These deposits are commonly located in the perforation or tubing. The injection of cold treating fluid promotes the formation of organic deposit such as petroleum, paraffins or asphaltenes. The deprivation of such organic deposit can be performed through various biodegradable bacteria that are prevalent in nature and may use these substances as a source of carbon and energy. Some bacteria can absorb specific alkanes, while others break down aromatic or resin fractions of hydrocarbons. Studies have identified bacteria from more than 79 genera that are capable of degrading organic deposit of oil and gas industries; several of these bacteria such as *Achromobacter, Acinetobacter, Alteromonas,*

Smita Lata and Swati Priya are the authors are equal contributed.

S. Lata · S. Priya · S. Mondal · S. Malik (✉) · J. Bora (✉)
Amity Institute of Biotechnology, Amity University Jharkhand, Ranchi, India
e-mail: smalik@rnc.amity.edu

J. Bora
e-mail: jbora@rnc.amity.edu

I. Barman
Program of Biochemistry, Faculty of Science, Assam down town University, Guwahati, Assam, India

S. Banerjee et al. (eds.), *Environmental Friendly Green Technologies for Improvement of Heavy Crude Oil Flow Assurance*, Environmental Science and Engineering,
https://doi.org/10.1007/978-3-031-86701-9_5

Arthrobacter, Burkolderia, Dietzia, Enterobacter, Mycobacterium, Pandoraea, Pseudomonas, Staphylococcus, Streptobacillus. Biodegradation of organic deposit occur under anaerobic and aerobic conditions.

5.1 Introduction

The utilization of petroleum products has experienced a substantial surge over the past few decades. Petroleum products primarily originate from crude oil and serve as the most extensively utilized primary energy resources of biological nature on a global scale. Developing nations characterized by inadequate environmental law enforcement are confronted with a significant challenge pertaining to oil leakage, particularly within marine ecosystems. The phenomenon of environmental contamination is experiencing a steady escalation because of improper practices involving the disposal and discharge of old engine oil into the open ocean. This often leads to significant disruption of ecosystems, affecting both the biotic and abiotic components of the environment. Leaks and accidental spills are a recurring concern that can arise at several stages of the petroleum industry, including exploration, production, refining, transportation, and storage of petroleum and its products. The presence of hydrocarbon contaminations poses significant risks to both plant health and human and animal well-being due to their potential carcinogenic, mutagenic, and immunotoxin effects. Soil contamination resulting from the presence of hydrocarbons has significant implications for the local ecosystem, as the accumulation of pollutants in animals and plant tissue can lead to mortality or genetic abnormalities. Consequently, the potential for contamination is consistently present in oil exploitation activities, particularly in challenging or distinctive ecosystems such as polar regions, deep sea areas, deserts, and wetlands. Moreover, the capacity to effectively address oil-contaminated surroundings in these settings is often inadequate. Oil pollution presents a challenging issue in terms of remediation. However, the presence of naturally occurring organic deposits in the environment has led to the evolution of bacteria that can decompose these organic deposits. These organisms have the potential to be utilized as viable options for the remediation of organic pollution. Consequently, bacteria have been subjected to screening processes and employed for the purpose of decomposing waste materials generated by various industries such as food, agriculture, chemicals, and pharmaceuticals. In recent years, the utilization of bacteria for the mitigation of environmental contaminants has emerged as a promising technique due to its cost-effectiveness and environmentally sustainable characteristics (Guerra et al. 2018).

Contaminated regions have a high abundance of indigenous microorganisms that can participate in the process of biodegradation. According to HAMzAH, the utilization of biodegradations offers a viable and productive approach for expediting the remediation of polluted environments (Hamzah et al. 2011). According to Geetha, microorganisms that are indigenous to certain contaminated areas have a significant role in the biodegradation of oil under favourable conditions (Geetha et al. 2013).

Several bacterial isolates could breakdown hydrocarbons, hence serving as a potential source of carbon and energy. The biological strategies are more cost-effective than physical and chemical methods for reducing hydrocarbon levels.

The ongoing progress and enhancement of microbial remediation technologies have also introduced a novel approach for addressing organic deposit pollution, garnering significant interest in the scientific community. The objective of this chapter is to offer recommendations for the future advancement of bacterial remediation techniques for organic deposits.

5.2 Toxic Impact of Organic Deposit of Oil and Gas Industry

The detrimental effects of oil pollution on the ecological environment have been widely acknowledge. The Deep-Water Horizon oil spill incident in the Gulf of Mexico had significant implications for both the economy and environmental security and continues to be a subject of considerable public interest. While there is growing public concern regarding the harmful consequences of oil pollution on human and animal populations in affected regions, the detrimental effects of organic deposits on microbial communities in these areas are frequently disregarded. According to the findings of Labud, microbial biomass was shown to be hindered by organic deposits, with the most significant detrimental impacts recorded in sandy soil contaminated with gasoline (Labud et al. 2007). In the conducted studies on diesel exposure, the researchers observed that the main consequences of diesel fuel toxicity were a decrease in species richness, evenness, and phylogenetic diversity. Consequently, the community composition was predominantly characterized by a small number of species, primarily *Pseudomonas*. Furthermore, van Dorst discovered a correlation between the decrease in species richness and phylogenetic diversity and the disturbance of the nitrogen cycle (Dorst et al. 2014). Specifically, they observed a significant reduction in species and functional genes associated with nitrification. The potential toxicity of naphthalene, 1-methylnaphthalene, and 2-methylnaphthalene, along with their oxygenated derivatives, on bacterial cells of *Agmenellum quadruplicatum*. They observed that these compounds did not exhibit any notable inhibitory effects on the growth of the bacterial cells. Nevertheless, the bacterial growth was impeded by the phenolic and quinone naphthalene derivatives. The observed phenomenon can be attributed to the increased solubility of phenols and quinones, which facilitates the transfer of molecules to bacterial cells. Consequently, these compounds exhibit more toxicity compared to the substances. Multiple studies have also documented that specific metabolic intermediate, which exhibit relatively high solubility and are generated during the decomposition of organic deposits by bacteria, may possess more cytotoxicity compared to the original molecules. Consequently, these intermediates have the potential to harm the bacteria. Nevertheless, it is worth noting that indigenous bacteria exhibit a tendency to form quite large aggregates, with each individual species serving a distinct and specific purpose. Consequently, certain bacteria that exhibit susceptibility to organic deposits experience significant inhibition when exposed to such compounds. Conversely, other bacteria that possess the ability to effectively breakdown organic deposits, together with those capable of exploiting cytotoxic intermediate intermediates, will thrive under these conditions. Nonetheless, the remediation of petroleum oil contaminants solely through the utilization of indigenous microorganisms is a time-consuming endeavour. Consequently, it becomes imperative to devise intervention strategies that can expedite this process.

5.3 Environmental Constraints

Various environmental parameters, including temperature, nutrient availability, electron acceptors, and substrates, are known to significantly impact bioremediation processes and exert influence on biodegradation reactions. The discrepancy between the performance of petroleum hydrocarbon-degrading bacteria in laboratory settings and field-scale experiments has been observed by numerous researchers. While these bacteria have demonstrated impressive degradation capabilities in controlled laboratory environments, their effectiveness in real-world scenarios has been found to be less desirable. The bacterial strains Acinetobacter sp. JLS1 and Pseudomonas aeruginosa JLC1, which were obtained from the Momoge wetlands in Jilin Province, China, exhibited varying degrees of sensitivity to temperature during the biodegradation of C16 alkane. This observation indicates that temperature exerts a significant influence on the efficiency of biodegradation (Xu et al. 2017). Effective degradation of petroleum hydrocarbons, specifically phenanthrene and dibenzothiophenes. However, the field experiment did not yield similar degradation effects, potentially due to variations in temperature conditions throughout the study. Temperature has been found to have an impact on bacterial proliferation and metabolism, as well as the soil matrix and the way pollutants are present. Consequently, this can indirectly influence the effectiveness of biodegradation. The provision of carbon, hydrogen, oxygen, nitrogen, sulfur, phosphorus, and several trace elements are widely recognized as essential for bacterial growth. Nevertheless, it is important for the environment to possess an adequate supply of additional nutritional elements to facilitate the growth of bacterial degraders, as petroleum hydrocarbons primarily consist solely of carbon and hydrogen. The conversion of 1 kg of hydrocarbons in bacterial cells requires the consumption of around 150 g of nitrogen and 30 g of phosphorous. Considerable attention has been dedicated to addressing this issue through comprehensive laboratory and field investigations. The successful application of fertilizers containing bioavailable nitrogen and phosphorus has been observed in stimulating the biodegradation of petroleum oil on various shorelines and sandy beaches. The presence of both soluble and non-soluble nutrients poses challenges in the process of remediation, resulting in reduced bioremediation efficiency. The utilization of nitrogen-fixing hydrocarbon-degrading bacteria as an alternative approach to provide nitrogen sources can enhance the effectiveness of bioremediation. The utilization of oxygen as an electron acceptor is of significant importance in aerobic degradation processes. However, its effectiveness is often compromised in environments contaminated with petroleum oil due to the restricted air permeability. A significant reduction of up to 75% in hydrocarbon pollutants was observed within a span of one year during field tests. This reduction was achieved by the implementation of aeration management and regulation techniques. Nevertheless, the cost and feasibility of ensuring an adequate oxygen supply for the purpose of promoting the bioremediation of petroleum contaminants in the environment are significant challenges. Therefore, the utilization of bulking agents, such as sawdust, in soil to enhance permeability, or the introduction of electron acceptors (such as NO_3^-, Fe^{3+}, or Mn^{2+}) into oxygen-deprived environments to stimulate anaerobic microorganisms, is frequently a more cost-effective approach compared to the provision of oxygen.

5.4 Hydrocarbon and Their Degrading Bacteria

Based on their chemical structure, hydrocarbons can be classified as follows: heteroatomic compounds (saturated and aromatic ones), such as resins and asphaltenes; aliphatic hydrocarbons, also known as saturated hydrocarbons; and aromatic hydrocarbons, which include both monocyclic and polycyclic aromatic hydrocarbons. Paraffins, often referred to as aliphatic hydrocarbons (AH), are mostly found in natural gas and oil deposits that are the result of plant and animal breakdown. Aliphatic hydrocarbons, with their general formula of C_nH_{2n}, are the most abundant elements of crude oil. They lack double bonds. They are weakly reactive at ambient temperature and have minimal water solubility due to their strong apolar nature caused by the lack of functional groups. They are divided into two groups based on their structures: cycloalkanes, which have a condensed ring structure, and alkanes, which have a linear structure and can also be branched. Different aliphatic hydrocarbons are typically used to substitute aromatic hydrocarbons, which have one or more benzene rings. They are primarily separated into two groups: polycyclic aromatic hydrocarbons (PAH) and monocyclic aromatic hydrocarbons (MAH), which together make up BTEX and include benzene, toluene, ethylbenzene, and xylene. The term "PAHs" refers to a class of lipophilic organic pollutants that are either products of incomplete combustion from anthropic (such as automobile emissions, home heating, and cigarette smoke) or biological sources (such as forest fires and bushes). They are substances of environmental significance because of their pervasiveness, persistence, and toxicity to both people and biota in the air, water, and soil.

A highly specialised kind of bacteria called methanotrophic ones can break down methane, the simplest hydrocarbon, and use it as a source of energy and carbon. On the other hand, these bacteria cannot proliferate on hydrocarbons with more carbon atoms. Due to its high hydrophobicity, low water solubility, and propensity to sorb to soil organic carbon, the complex high-molecular-weight pyrene has a low bioavailability and is resistant to biodegradation. The first reports of pyrene degradation date back to 1988, and they concerned a strain of Mycobacterium that could mineralise pyrene and use it as its only source of carbon and energy. Later, a number of pyrene degraders have also been found; these include members of the genera Mycobacterium, Sphingomonas, Rhodococcus, Bacillus, Burkholderia, Cycloclasticus, Pseudomonas, and Stenotrophomonas. The percentages of these degraders in soil vary from 32 to 96%, and their degradation durations range from 4 to 42 days. Longer aliphatic chains and aromatic molecules persist after the volatile hydrocarbon portion quickly evaporates when significant amounts of oil (hydrocarbon mixture) are released into the environment. Oil floats on the surface and leaves blotches because it is less thick than water and insoluble. Hydrocarbon-oxidizing bacteria require insoluble petroleum droplets to stick to, as these exhibit a high cell count. These organisms' actions cause oil to deteriorate and oily areas to collapse. Longer aliphatic chains and aromatic molecules persist after the volatile hydrocarbon portion quickly evaporates when significant amounts of oil (hydrocarbon mixture) are released into the environment. Oil floats on the surface and leaves blotches because it is less thick than water and insoluble. Hydrocarbon-oxidizing bacteria require insoluble petroleum droplets to stick to, as these exhibit a high cell count. These organisms' actions cause oil to deteriorate and oily areas to collapse. Several investigations have demonstrated that bacterial communities in contact with hydrocarbons quickly transform into hydrocarbonoclastic bacteria, which are able to break down and use hydrocarbon molecules as sources of carbon and energy. In order to survive in the presence of hydrocarbons, hydrocarbonoclastic bacteria have developed adaptive mechanisms. These include the capacity to emulsify and metabolise hydrocarbons, the activation of DNA repairing mechanisms, the synthesis of molecules involved in quorum sensing and biofilm mechanisms, and the control of efflux pumps and pores to regulate the concentration of hydrocarbons inside a cell. Adhesion to the substrate and the synthesis of substances such biosurfactants/bioemulsionants, biopolymers, solvents, gases, and acids to make the hydrocarbon accessible are the processes involved in the biodegradation of a hydrocarbon. Specifically, biosurfactants are amphiphilic compounds with varying molecular weights that can be produced by a variety of microbes. The synthesis and function of these compounds are actually closely related to the mechanisms of absorption (uptake) of such substances inside bacterial cells. In prokaryotes, the ability to produce these compounds is often matched with the ability to grow on insoluble substrates, such as hydrocarbons. These compounds frequently have the ability to detoxify the substrate. According to research by Posada-Baquero, the presence of rhamnolipid, a biosurfactant, in the sunflower rhizosphere promoted the desorption of PAHs in soil contaminated with creosote. Furthermore, in a slurry biodegradation experiment, the acceleration of the sluggish desorption of PAHs led to faster biodegradation. Using cultures enhanced with phenanthrene, a Pseudomonas strain (W10) from a diesel-contaminated soil. In the presence of a wide range of hydrocarbons, such as aliphatic, monocyclic aromatic, and polycyclic aromatic hydrocarbons, this strain demonstrated prospective growth. This was brought about by the strain's capacity to produce a biosurfactant, which increased the hydrocarbons' availability for breakdown. Pseudomonas aeruginosa was found to be able to break down polycyclic aromatic hydrocarbons (PAHs) such pyrene, phenanthrene, and fluorene as well as n-alkanes (C16 and C19) in another study. Strains of Enterobacter cloacae that exhibit a high capacity to emulsify hydrocarbons, lower surface tension, and accelerate the degradation of these pollutants have been isolated by other authors. Hydrocarbon degradation has also been linked to the genus *Delftia. Delftia sp. NL1*, for instance, was isolated from an Algerian oil field and degraded over 66.76% of diesel in just 7 days. This process produced glycolipids, which functioned as biosurfactants and enhanced the insoluble hydrocarbon fraction's accessibility and bioavailability. The *Achromobacter (AC15)* bacteria, which was identified, was another generator of biosurfactants and could only utilise high pyrene concentrations as a source of carbon and energy. Because it generated bioemulsionants that could lower the surface tension of the culture medium (from 67.2 to 33.2 mN/m), it was able to breakdown around 40% of the pyrene provided at an initial concentration of 300 mg/L in just 14 days. According to a recent study that showed an increase in their relative abundance in the presence of diesel, heavy metals, and PAHs, bacteria in the genus Acinetobacter also play a significant role in the biodegradation of PAHs. The Acinetobacter genus is well recognised for its effectiveness in breaking down several hydrocarbons, including naphthalene, acenaphthene, and acenaphthylene, as well as monoaromatic chemicals. Hydrocarbons are recognised to be the only carbon and energy source that the Stenotrophomonas genus can utilise. From a soil contaminated with crude hydrocarbons, Elufisan. identified Stenotrophomonas sp. Pemsol and demonstrated the organism's ability to develop when five distinct PAHs (biphenyl, anthraquinone, phenanthrene, naphthalene, and phenanthridine) were present as the only sources of carbon and energy. This sequenced strain's whole genome showed that 145 genes involved in PAH breakdown were present. The same scientists also noted that other bacteria can acquire certain genes related to the catalytic metabolism of hydrocarbons through horizontal gene transfer. Table 5.1 provides an overview of all the bacterial taxa that were involved in the hydrocarbon biodegradation process in the research mentioned above.

5.5 Metabolic Restriction in Degradation of Organic Deposits by Bacteria

The ability to biodegrade petroleum oil is associated with the concentration and composition of hydrocarbons (Fig. 5.1). Extremely high levels of petroleum hydrocarbons strongly inhibit bacterial growth, resulting in poor biodegradation efficiency and even death of the bacteria. The order of biodegradability of hydrocarbons is as follows: linear alkanes > branched alkanes > low molecular weight alkyl aromatics > monoaromatics > cyclic alkanes > polyaromatics > asphaltenes. This is related to the physico-chemical properties of the substrate and its bioavailability, which affect the contact, transport, and transformation of bacteria to hydrocarbon substrates. The vast

Table 5.1 Example of single bacteria involved in hydrocarbon degradation

Genus/species	Mechanism of action	Hydrocarbon	References
Proteobacteria			
Achromobacter (AC15)	Biosurfactant production	Pyrene	Pandolfo et al. (2023)
Acinetobacter	Biosurfactant production	naphthalene, acenaphthene, acenaphthylene	
Delftia sp. NL1	Biosurfactant production	Diesel	
Enterobacter cloacae	Biosurfactant production	Crude oil	
Pseudomonas aeruginosa	Biosurfactant production	n-alkanes (C16–C19), fluorene, phenanthrene, pyrene	
Pseudomonas putida	Chemotaxis	Pyrene	
Pseudomonas W10	Biosurfactant production	Phenanthrene	
Stenotrophomonas sp. Pemsol	Horizontal gene transfer	Biphenyl, anthraquinone, phenanthrene, naphthalene, phenanthridine	
Actinobacteria			
Rhodococcus sp. P14	Change in fatty acid composition of cell membrane	Biofilm formation/ phenanthrene, pyrene, benzo(a)pyrene	Song et al. (2011)

majority of indoor studies are focused on the degradation of a single substrate, but in nature the components of petroleum hydrocarbon pollutants are extremely complex. Accordingly, it is difficult to reproduce laboratory results in practical applications. For example, *Pseudomonas putida F1* can efficiently mineralize benzene, toluene, and phenol. While in the substrate mixtures, toluene and benzene enhance the biodegradation of phenol; however, phenol inhibits the biodegradation of benzene and toluene. The key components of bacterial degradation of petroleum hydrocarbons are various specific enzymes. For example, the enzymes alkane 1-monooxygenase, alcohol dehydrogenase, cyclohexanoldehydrogenase, methane monooxygenase and cyclohexanone 1,2 monooxygenase are involved in degradation of alkanes, while naphthalene 1,2-dioxygenase ferredoxin reductase component,cis-2,3-dihydrobiphenyl-2,3-diol dehydrogenase and salicylaldehyde dehydrogenase are associated with naphthalene degradation and benzene dioxygenase, toluene dioxygenase and ethylbenzene dioxygenase work on other petroleum hydrocarbons (Bacosa et al. 2018). Many isolated bacteria possess the ability to mineralize chemically simple petroleum hydrocarbons completely, such as linear alkanes, if these bacteria possess all of the enzymes for the targeted substrate. However, few bacteria can completely mineralize complex compounds such as resins and asphaltenes because of the lack of some enzymes. The advantages of microbial communities are presented because there are a variety of catabolic genes in a bacterial consortium, and the synergistic effects of these genes are beneficial to achieving the purification of pollutants. A bacterial consortium composed of five culturable bacteria has been constructed by Wanapaisan et al. (2018). Researchers found that these five bacteria showed synergistic pyrene degradation due to the following aspects: (1) The *Bacillus* strain enhanced the bioavailability of the pyrene by producing biosurfactant, (2) two *Mycobacterium* strains contributed to the initiation of pyrene degradation, and (3) *Novosphingobium* and *Ochrobactrum* efficiently degraded the intermediates of pyrene. Given the complexity of the petroleum components, construction of the minimal functional bacterial consortium or genetic engineering bacteria for bioremediation of petroleum oil has become a trend in this field (Dvořák et al. 2017). However, the stability of the community and the safety of the engineered bacteria are additional problems that must be overcome.

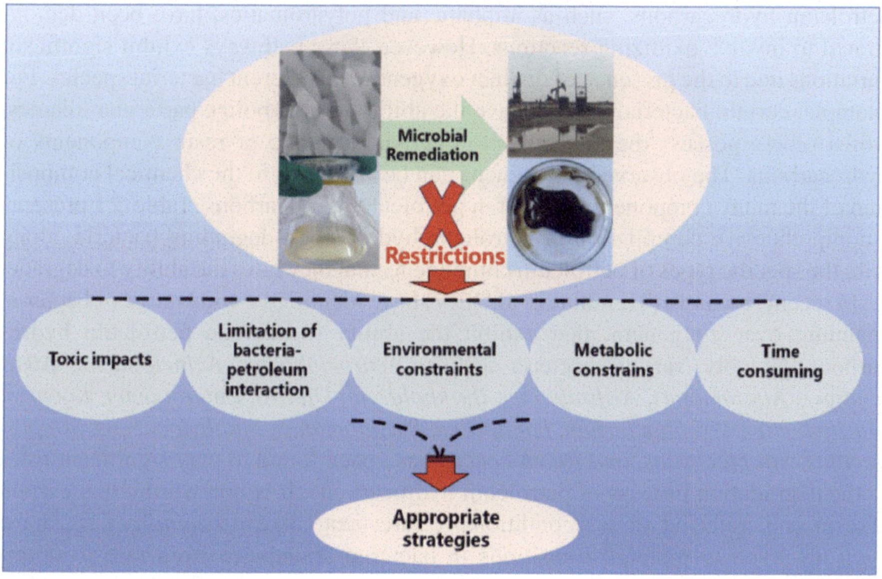

Fig. 5.1 Metabolic restriction in degrading of organic deposits. (Frontiers in microbiology|www.frontiersin.org 6 December 2018|Volume 9|Article 2885

5.6 Petroleum Hydrocarbon-Degrading Bacteria and Their Preferred Degradation Substrates

The degradation or metabolism of petroleum hydrocarbons in the environment is primarily carried out by native bacteria. These bacteria utilize the hydrocarbons as a source of energy and carbon for their growth and reproduction. Additionally, the presence of petroleum hydrocarbons in the microbial environment can cause physiological stress, which further motivates the bacteria to alleviate this stress through degradation or metabolism. The utilization of microbial biotechnology and high throughput sequencing technologies, including microfluidic approaches (Guerra et al. 2018), has shown advantageous in the process of screening and detecting functional bacteria within settings contaminated by petroleum hydrocarbons. Numerous studies have provided evidence indicating the presence of a significant population of hydrocarbon-degrading bacteria in environments characterized by high concentrations of oil, such as areas affected by oil spills and oil reservoirs. Moreover, these studies have demonstrated a strong correlation between the abundance and quantity of these bacteria and the specific types of petroleum hydrocarbons present, as well as the surrounding environmental conditions. A wide range of both common and specialized bacterial species have been identified and employed as bio degraders in the treatment of petroleum hydrocarbons. The degradation pathways of several petroleum hydrocarbons, such as aliphatic and polyaromatics, have been demonstrated to involve oxidizing reactions. However, these pathways exhibit significant variations due to the presence of distinct oxygenase in different bacterial species. For example, certain bacterial species have the ability to metabolize particular alkanes, whilst others possess the capability to degrade aromatic or resin components of hydrocarbons. The observed phenomena can be attributed to the chemical composition of the many components present in petroleum hydrocarbons. Table 5.1 presents a comprehensive compilation of petroleum hydrocarbon-degrading bacteria, along with the specific types of petroleum components that they have the ability to degrade.

In recent research, (Tremblay et al. 2017) have identified a wide range of bacteria, spanning over 79 genera, that exhibit the ability to degrade petroleum hydrocarbons. Notably, various bacteria such as *Achromobacter, Acinetobacter, Alkanindiges, Alteromonas, Arthrobacter, Burkholderia, Dietzia, Enterobacter, Kocuria, Marinobacter, Mycobacterium, Pandoraea, Pseudomonas, Staphylococcus, Streptobacillus, Streptococcus, and Rhodococcus* have been found to play significant roles in the degradation process of petroleum hydrocarbons. It is noteworthy that certain taxa in soil, referred to as "conditionally rare taxa," like *Alkanindiges* sp., have been observed to undergo transitions in bacterial abundance from rare to dominant states, which are significantly influenced by environmental factors such as diesel pollution. In a study conducted by Yakimov et al. (2003), it was observed that certain obligate hydrocarbonoclastic bacteria (OHCB), such as *Alcanivorax, Marinobacter, Thallassolituus, Cycloclasticus, Oleispira,* and a few others, exhibited a minimal presence or were undetectable prior to pollution. However, following petroleum oil contamination, these OHCB species were found to be prevalent. These observations imply that these microorganisms play a vital part in the breakdown of petroleum hydrocarbons, and they have a substantial impact on the alteration and ultimate destiny of petroleum hydrocarbons within the surrounding ecosystem. Whereas several bacteria have a wide range of capabilities in degrading petroleum hydrocarbons, *Dietzia sp.* DQ12-45-1b specifically utilizes n-alkanes (C6–C40) and various other compounds as their exclusive sources of carbon. *Achromobacter xylosoxidans DN002* demonstrates effective degradation capabilities for a range of monoaromatic and polyaromatic hydrocarbons. However, it is important to note that the complete degradation of the full petroleum hydrocarbon component remains a challenge for most bacterial species. It has been shown that the ability of bacteria to degrade or consume petroleum hydrocarbon components is limited to specific types, rendering certain components inaccessible to them. The variation in roles of indigenous bacteria in oil-contaminated locations can be related to the presence of different catalytic enzymes among them. Furthermore, it can be inferred from the statement that the effective remediation of petroleum hydrocarbon contamination necessitates the collaborative efforts of several functional bacteria in order to attain optimal environmental purification outcomes. A halotolerant Hydrocarbon Utilizing Bacterial Consortium (HUBC) was developed, comprising of the bacterial isolates *Ochrobactrum sp., Stenotrophomonas maltophilia,* and *Pseudomonas aeruginosa.* This consortium demonstrated effective degradation of crude oil at a concentration of 3% v/v, achieving a degradation percentage of 83.49%. It was observed that a specific co-culture consisting of an indigenous bacterial consortium and exogenous Bacillus subtilis to successfully expedite the process of crude oil breakdown. It was observed that an aboriginal bacterial consortium, originating from the Penglai 19–3 oil spill incident in China, exhibited superior oil degradation capabilities when compared to individual bacterial strains. The findings also indicated that this indigenous consortium displayed promising potential for the bioremediation of crude oil dispersed within the marine ecosystem. According to a study conducted in the field, the introduction of a synthetic consortium consisting of *Aeromonas hydrophila, Alcaligenes xylosoxidans, Gordonia sp., Pseudomonas fluorescens, Pseudomonas putida, Rhodococcus equi, S. maltophilia,* and *Xanthomonas sp.* resulted in a significant enhancement of biodegradation efficiency, reaching 89%, during a 365-day treatment of soil contaminated with diesel oil. Some results are summarized in Table 5.2. Collectively, these results suggest that enhancing the ability of microorganisms to break down petroleum hydrocarbons by utilizing bacterial consortia with various catabolic genes is a viable and practical approach to expedite the removal of these contaminants from polluted ecosystems. The physical contact between bacteria and petroleum hydrocarbons can be explained using Fig. 5.2.

Table 5.2 Petroleum hydrocarbon degrading bacteria and their preferred degradation substrates

Petroleum hydrocarbon components	Bacterial species	Main degradation profile	Reference
Aromatics	*Achromobacter xylosoxidans* *Aeribacillus pallidus* *Mycobacterium cosmeticum* *Pseudomonas aseruginosa* *Cycloclasticus* *Neptunomonas napthovoran* *Bacillus Licheniformis* *Bacillus Mojavensis Sphingomonas, Sphingobium and Novosphingobium*	Mono-/polyaromatics Mono-/polyaromatics Monoaromatics Monoaromatics Polyaromatics Polyaromatics Polyaromatics Polyaromatics	Varjani (2017)
Resins and asphaltenes	*Pseudomonas sp.* *Pseudomonas spp.* *Bacillus sp.* *Citrobacter sp.* *Enterobacter sp.* *Staphylococcus sp.* *Lysinibacillus sp.* *Bacillus sp.* *Pseudomonas sp.*	Resins Asphaltenes Asphalteness	
Aliphatics	*Dietzia sp.* *Pseudomonas sp.* *Oleispira antarctica* *Rhodococcus ruber* *Geobacillus thermodenitrifican* *Rhodococcus sp.* *Alcanivorax sp.* *Gordonia sihwensis*	n-alkanes(C6-C40) n-alkanes (C14-C30) n-alkanes (C-10-C18) n-alkanes (C13-C17) n-alkanes (C15-C36) Cyclohexane n-alkanes and branched alkanes Branched and normal alkanes	

5.7 Discussion

Over the past few decades, a significant number of hazardous waste sites have emerged on a global scale due to the accumulation of xenobiotics in soil, water, and air. Xenobiotic substances have been found to contaminate the environment due to their resistant nature. One example of contamination is petroleum (hydrocarbon) pollution. Numerous microorganisms have already been identified and extracted from two primary environments, namely soil and marine sources, which are predominantly affected by petroleum hydrocarbon contamination.

In the present investigation, the bacterial strains were extracted from soil specimens that had been subjected to contamination by petroleum hydrocarbons. The diesel fuel and burned engine oil as the exclusive carbon sources to assess the degrading effectiveness of the isolated bacteria in their experimental setup. Although these isolates have the capability to degrade diesel oil and burnt engine oil, their

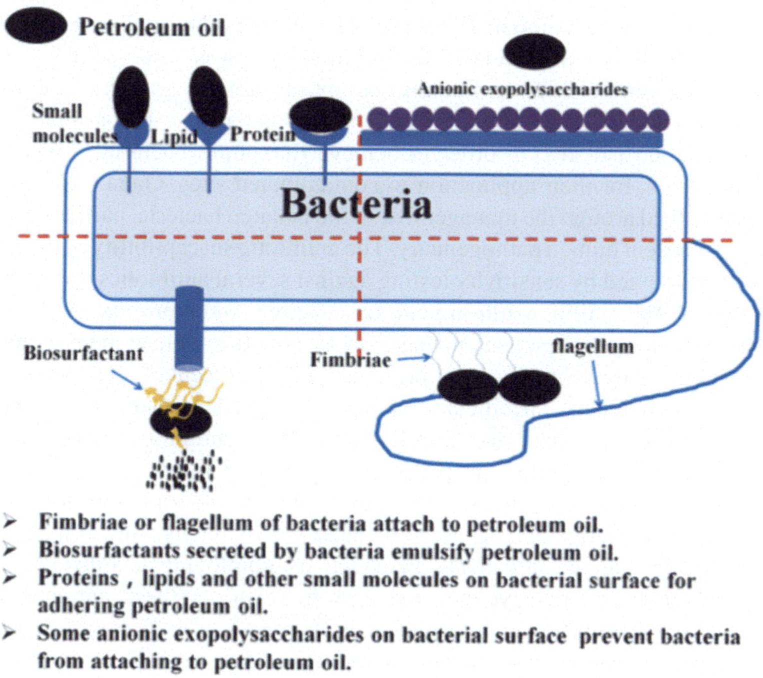

Fig. 5.2 Schematic diagram of the physical contact between bacteria and petroleum hydrocarbons. (Frontiers in microbiology|www.frontiersin.org 6 December 2018|Volume 9|Article 2885 https://doi.org/10.3389/fmicb.2018.02885

ability to do so is contingent upon the prevailing physical, chemical, and biological circumstances.

In order to determine the identity of these isolates, the most reliable genetic method is to compare the sequences of the bacterial 16S rRNA gene. The analysis of the 16S rRNA gene sequencing data revealed the presence of Pseudomonas sp., Acinetobacter sp., and Enterobacter sp. as the identified bacterial isolates. The isolates were subjected to morphological and biochemical testing in order to characterize them.

The morphological test findings reveal that all of the bacteria that were isolated exhibit characteristics of being gram negative, having a cocci form, and displaying motility. The biochemical findings suggest that the isolates possess the ability to detoxify hydrogen peroxide and also have the capacity to utilize citrate as a carbon source. The sugar utilization test additionally verified their ability to metabolize various carbohydrates, including glucose, sucrose, and lactose, as nutritional substrates. All of the isolates exhibit the ability to ferment lactose.

Furthermore, these organisms possess the capability to undergo hydrolysis of urea and starch. It was observed that all of the examined isolates exhibited the ability to thrive in MS media that was enriched with diesel oil and burned engine oil, serving as the sole carbon source. The growth factors, such as pH and temperature, were tuned by the researchers to achieve maximal growth of the isolates. The isolates exhibited the greatest growth rate at a pH level of 7.0. Additionally, it was observed that a temperature of 37 degrees Celsius was found to be optimal for the growth of the isolated bacteria. The objective of the optimization was to determine the optimal growing conditions for the isolates, in order to achieve maximum bacterial growth while minimizing costs, for their application in contaminated sites. One of their primary concerns revolved around the management of the isolated bacteria, particularly in the event of inadvertent human pathogenicity. The antibiotic susceptibility of the isolated bacteria was assessed by sensitivity testing against several antibiotics. Levofloxacin, doxycycline, carbenicillin, azithromycin, tetracycline, and ciprofloxacin have been employed as therapeutic agents to manage severe situations due to their notable efficacy, as evidenced by their consistent presence in 0.01 percent of all isolates. One of the three microorganisms, specifically *Enterobacter sp.*, demonstrated susceptibility to levofloxacin, tetracycline, and ciprofloxacin. These antibiotics have the potential to effectively manage human pathogenicity caused by this bacterium. When dealing with *Acinetobacter sp.*, levofloxacin, azithromycin, and ciprofloxacin have been employed as therapeutic agents to manage pathogenicity. Nevertheless, *Pseudomonas* species can be effectively managed by employing levofloxacin, doxycycline, azithromycin, tetracycline, and ciprofloxacin, as these antibiotics have demonstrated susceptibility against this bacterium.

This chapter examines the efficacy of isolates in the removal of diesel oil and burned engine oil. All the isolates exhibited the greatest degrading capacity in a 4% (v/v) concentration of diesel oil and in an 8% concentration of burned engine oil in the MS broth medium. In the present investigation, it was revealed that the growth rates of the isolates exhibited a decline as the concentration of diesel oil increased. On the contrary, it was observed that the growth rates of bacteria exhibited an upward trend as the concentration of burned engine oil was increased. Their isolates exhibit a greater capacity for degrading burned engine oil compared to diesel oil. The study's notable findings indicate that the accumulation and contamination of burned engine oil, which is not being further utilized, can be effectively mitigated from the environment through the application of these isolates. The biodegradation process by microorganisms can result in the degradation of less than 4% of diesel oil. Remarkably, the isolates in question exhibit a higher capacity for the degradation of diesel oil and burned engine oil as compared to the findings presented in this paper.

All three isolates represent the most potent naturally existing bacteria capable of decomposing oil. This study demonstrates the methodology employed to isolate oil-degrading bacteria from a location contaminated with oil. The findings suggest that oil-affected soil serves as the principal reservoir for the separation of petroleum-degrading bacteria. It is strongly recommended that all these isolates have the potential to be utilized for bioremediation applications in the removal of oil from contaminated regions.

5.8 Conclusion

Organic deposit poses a significant concern as pollutants, primarily due to their pronounced toxicity towards both human and environmental well-being. The utilization of organic deposits degrading bacteria for bioremediation is widely recognized as a sustainable and effective technological approach. Numerous bacterial species possessing the capacity to degrade organic deposits have been harnessed and utilized in the field of bioremediation. Nevertheless, some obstacles that impede the rate of biodegradation have been identified during practical implementation. This chapter elucidated various limiting variables, such as the adverse impacts of organic deposits, the bioavailability of contaminants, environmental limitations, metabolic constraints, and temporal demands. Subsequently, it provided a comprehensive overview of the existing strategies employed to address these challenges. Numerous solutions, including the regulation of environmental conditions and the optimization of microbial inoculants, have been thoroughly examined and successfully implemented. Based on the existing body of literature examined in this study, it is evident that further research endeavours are necessary before the effective implementation of bioremediation can be achieved for the purpose of rehabilitating habitats contaminated with petroleum oil. This study also demonstrates the isolation of oil-degrading bacteria from soil contaminated with oil, highlighting their potential for use in bioremediation efforts. This discovery implies that the utilization of these isolates presents a viable treatment option for the removal of oil from polluted soil or water through the process of biodegradation.

References

Bacosa HP et al (2018) Hydrocarbon degradation and response of seafloor sediment bacterial community in the northern Gulf of Mexico to light Louisiana sweet crude oil. ISME J 12(10):2532–2543

Dvořák P, Nikel PI, Damborský J, de Lorenzo V (2017) Bioremediation 3.0: engineering pollutant-removing bacteria in the times of systemic biology. Biotechnol Adv 35(7):845–866

Geetha S, Joshi SJ, Kathrotiya S (2013) Isolation and characterization of hydrocarbon degrading bacterial isolate from oil contaminated sites. APCBEE Proc 5:237–241

Guerra AB et al (2018) Metagenome enrichment approach used for selection of oil-degrading bacteria consortia for drill cutting residue bioremediation. Environ Pollut 235:869–880

Hamzah AIO, Tavakoli A, Rabu A (2011) Detection of toluene degradation in bacteria isolated from oil contaminated soils. Sains Malays 40(11):1231–1235

Labud V, Garcia C, Hernandez T (2007) Effect of hydrocarbon pollution on the microbial properties of a sandy and a clay soil. Chemosphere 66(10):1863–1871

Pandolfo E, Barra Caracciolo A, Rolando L (2023) Recent advances in bacterial degradation of hydrocarbons. Water 15(2):375

Song X et al (2011) Isolation, characterization of Rhodococcus sp. P14 capable of degrading high-molecular-weight polycyclic aromatic hydrocarbons and aliphatic hydrocarbons. Mar Pollut Bull 62(10):2122–2128

Tremblay J et al (2017) Chemical dispersants enhance the activity of oil-and gas condensate-degrading marine bacteria. ISME J 11(12):2793–2808 van Dorst J et al (2014) Bacterial targets as potential indicators of diesel fuel toxicity in subantarctic soils. Appl Environ Microbiol 80(13):4021–4033

Varjani SJ (2017) Microbial degradation of petroleum hydrocarbons. Biores Technol 223:277–286

Wanapaisan P et al (2018) Synergistic degradation of pyrene by five culturable bacteria in a mangrove sediment-derived bacterial consortium. J Hazard Mater 342:561–570

Xu X et al (2017) Synergetic effect of bio-photocatalytic hybrid system: g-C3N4 and acinetobacter sp. JLS1 for enhanced degradation of C16 alkane. Chem Eng J 323:520–529

Yakimov MM et al (2003) Oleispira antarctica gen.nov., sp.nov., a novel hydrocarbonoclastic marine bacterium isolated from Antarctic coastal sea water. Int J Syst Evol Microbiol 53(3):779–785

Chapter 6
Potential Ecological Paraffin and Wax Degrading Bacterial Strains and Their Characterizations

Gourav Patil, Sankha Chakrabortty, and Shirsendu Banerjee

Abstract The exploration of potential ecological paraffin and wax-degrading bacterial strains has garnered significant attention within the field of environmental biotechnology due to the increasing need for sustainable solutions to environmental pollution. Paraffin and wax, both of which are hydrocarbons, are widely used in various industries, and their accumulation in the environment can lead to detrimental ecological impacts. These substances are often resistant to natural degradation, making them persistent pollutants in many ecological settings. This chapter discusses the characterization of selected bacterial strains that have demonstrated exceptional efficacy in hydrocarbon remediation processes, with a particular focus on paraffin and wax degradation. The bacteria's ability to break down these complex hydrocarbons is not only a testament to their unique metabolic pathways but also highlights their potential as bioremediation agents. By harnessing these bacterial strains, it is possible to mitigate the environmental burden of hydrocarbon pollutants. Understanding the mechanisms through which these bacterial strains degrade paraffin and wax is essential for optimizing their application in real-world scenarios. This includes studying the enzymes involved, the conditions that favor biodegradation, and the potential for scaling these processes for industrial use. Moreover, it is crucial to assess the ecological implications of paraffin and wax degradation. While these bacterial strains offer a promising solution to pollution, their introduction into natural environments must be carefully managed to avoid unintended ecological consequences.

G. Patil
School of Biotechnology, Kalinga Institute of Industrial Technology, Bhubaneswar, India

S. Chakrabortty · S. Banerjee (✉)
School of Chemical Engineering, Kalinga Institute of Industrial Technology, Bhubaneswar, India
e-mail: shirsendu.banerjee@kiitbiotech.ac.in

S. Banerjee et al. (eds.), *Environmental Friendly Green Technologies for Improvement of Heavy Crude Oil Flow Assurance*, Environmental Science and Engineering,
https://doi.org/10.1007/978-3-031-86701-9_6

6.1 Introduction

Paraffin and wax are intricate mixtures of hydrocarbons with diverse origins and compositions, making them significant subjects of study in both industrial applications and environmental management. Paraffin wax, which is predominantly derived from petroleum, consists mainly of saturated hydrocarbons, making it highly stable and resistant to degradation (Mansoori et al. 2003). In contrast, natural waxes, which are sourced from plants, animals, or even insects, contain a more complex mixture of esters, fatty acids, and hydrocarbons. These natural waxes serve various biological functions, such as protecting plants from water loss or insects from environmental stress, but their complex composition also makes them challenging to break down in natural settings. In the context of crude oil production and refining processes, paraffin and wax are closely linked to operational challenges. Crude oil is itself a complex mixture of hydrocarbons, and the presence of paraffin within it can lead to significant issues during transportation and storage. Under specific conditions, such as low temperatures or changes in pressure, paraffin can precipitate out of the crude oil, forming wax deposits (Misra et al. 1995). These deposits are notorious for causing blockages in pipelines, reducing the efficiency of oil extraction and transportation, and necessitating costly maintenance and intervention efforts. The oil industry has long sought effective solutions to manage and mitigate the formation of these wax deposits, as they pose substantial operational challenges and economic burdens (Theyab 2020).

Beyond their industrial significance, the environmental impacts of paraffin and wax residues are a growing concern. Accidental releases of these compounds, whether through spills during oil extraction and transportation or through improper disposal practices, can lead to severe contamination of soil and water bodies. Once in the environment, paraffin and wax can persist for long periods due to their resistance to natural degradation processes. In terrestrial ecosystems, these compounds can coat the surface of plants, forming a barrier that impedes critical physiological processes such as gas exchange and nutrient uptake. This interference can stunt plant growth, reduce agricultural yields, and disrupt ecosystem functioning, leading to cascading effects on biodiversity and ecological balance (Silva et al. 2022). In aquatic environments, the deposition of wax can be equally problematic. Wax layers can create physical barriers that alter the dynamics of habitats, affecting the movement, feeding, and reproduction of aquatic organisms. For example, fish, invertebrates, and other aquatic species may struggle to navigate through wax-coated surfaces or find their usual food sources, leading to declines in population and biodiversity. Moreover, the presence of wax in water bodies can affect the quality of the water itself, making it more difficult for ecosystems to sustain life (Feild et al. 2005; Hadley 1989). Given the environmental and operational challenges associated with paraffin and wax, there is a growing interest in identifying and utilizing microbial strains capable of degrading these compounds. Microbial degradation offers a promising solution for mitigating the environmental impact of paraffin and wax contamination. Certain bacteria and fungi have evolved the ability to break down hydrocarbons, including the complex mixtures found in paraffin and wax. By harnessing these microbial strains, it may be possible to develop bioremediation strategies that can effectively reduce the persistence of these pollutants in the environment, restoring the health and balance of affected ecosystems. This area of research holds significant potential not only for environmental cleanup but also for enhancing the sustainability of industrial processes by reducing the ecological footprint of hydrocarbon extraction and refining.

6.1.1 Need for Paraffin and Wax-Degrading Bacterial Strains

Recent advancements in bioremediation have shown promise in addressing paraffin and wax contamination. Microbial communities, particularly bacterial strains, have demonstrated the ability to degrade these hydrocarbons, offering a sustainable approach to mitigating environmental impact (Griffin et al. 2014). Bacterial strains capable of degrading paraffin and wax exhibit diverse enzymatic mechanisms and metabolic pathways. These microorganisms often possess specialized enzymes, such as alkane hydroxylases and wax esterases, which facilitate the breakdown of complex hydrocarbon structures (Wentzel et al. 2007). Moreover, some bacterial strains can utilize paraffin and wax compounds as carbon and energy sources through metabolic pathways like β-oxidation (Mohanan et al. 2020). Understanding the enzymatic capabilities and metabolic pathways of paraffin and wax-degrading bacterial strains is crucial for optimizing bioremediation strategies. By elucidating the mechanisms underlying microbial degradation, Efficiency and specificity of bioremediation processes can be enhanced, leading to more effective restoration of contaminated environments.

6.1.1.1 Paraffin and Wax-Degrading Bacterial Strains

Traditional methods for the removal of paraffin and wax contaminants, such as mechanical or chemical treatments, are often costly, labor-intensive, and may pose secondary environmental risks. In contrast, bioremediation offers a sustainable and environmentally friendly approach to mitigating hydrocarbon pollution (Azubuike et al. 2020). Among bioremediation techniques, microbial degradation stands out as a promising strategy due to its ability to target specific contaminants and operate under diverse environmental conditions (Ławniczak et al. 2020). Bacterial strains capable of degrading paraffin and wax compounds play a crucial role in bioremediation efforts. These microorganisms possess specialized metabolic pathways and enzymes that enable them to utilize hydrocarbons as carbon and energy sources (Wentzel et al. 2007). By harnessing the metabolic potential of these bacteria, it is possible to enhance the degradation of paraffin and wax residues in contaminated environments (Van Hamme Jonathan et al. 2003). Furthermore, the application of paraffin and wax-degrading bacterial strains offers several advantages over conventional remediation methods. Microbial degradation is often less invasive and can be applied in situ, minimizing disruption to ecosystems and reducing remediation costs (Ossai et al. 2020). Moreover, microbial communities are adaptable and can evolve to degrade complex hydrocarbon mixtures, making them versatile tools for environmental cleanup. Given the persistent nature of paraffin and wax contamination and the increasing demand for sustainable remediation solutions, there is a growing need to explore and utilize bacterial strains with the capability to degrade these hydrocarbon compounds effectively. Understanding the ecological roles and metabolic activities of these microorganisms is essential for developing targeted bioremediation strategies and restoring contaminated environments to their natural state.

Pseudomonas sp. Strain PW-1

The strain PW-1 exhibits a degradation rate of 52.85% for paraffin wax, indicating its ability to utilize 1050 mg of paraffin wax as the sole carbon source in a 1000 mL minimal salts medium. This highlights the potential utility of strain PW-1 in addressing paraffin deposition challenges in oil wells to enhance oil recovery. The exploration into the biodegradation of paraffin wax by *Pseudomonas* sp. PW-1 sheds light on critical factors influencing this process. A study by Zhang et al. shows temperature emerges as a pivotal variable, with optimal enzymatic activity observed within the 20–40 °C range, peaking at 30 °C. This underscores the strain's preference for moderate temperatures, aligning with its highest enzymatic activities for paraffin wax degradation. Moreover, nitrogen sources play an important role in alkane. Notably, $(NH_4)_2SO_4$ supplementation demonstrates superior efficacy in fostering both degradation and bacterial growth. The significance of the carbon-to-nitrogen ratio in degradation processes is evident, with the study showcasing optimal degradation rates at specific ratios. Paraffin wax, coupled with $(NH_4)_2SO_4$, elucidates the importance of a balanced carbon-to-nitrogen ratio, with a ratio of 10:1 yielding the highest degradation rate. Figure 6.1 shows the schematic of the test and results in brief. Practical tests conducted under controlled laboratory conditions further validate findings, with a notable degradation rate of 52.85% observed for paraffin wax. This underscores the potential application of *Pseudomonas* sp. PW-1 in environmental remediation efforts targeting hydrocarbon pollutants (Zhang et al. 2016).

Gordonia Species (*Gordonia amicalis* LH3)

Gordonia species, recognized for its hydrocarbon-degrading capabilities is a Isolated from an oil-producing well, it has demonstrated proficiency in breaking down alkanes, emphasizing its potential for environmental remediation. *Gordonia amicalis* LH3 stands out for its efficient utilization of long-chain n-alkanes as carbon and energy sources. Under aerobic conditions, this strain has been shown to exhibit robust

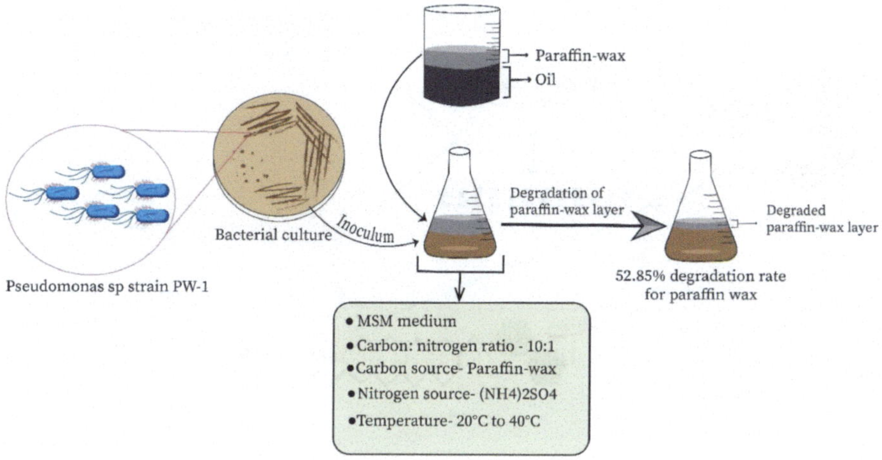

Fig. 6.1 Degradation of paraffin and wax by *Pseudomonas* sp. strain PW-1

growth on oil substrates, highlighting its potential for bioremediation applications. The facultative anaerobic bacterium, *G. amicalis* LH3, Under aerobic conditions, it demonstrates significant paraffin degradation, reducing it by 18.0% (w/w) at a rate of 36 mg/day. In contrast, under anaerobic conditions, the degradation of paraffin is less pronounced, at 2.3% (w/w) with a rate of 4.4 mg/day. Moreover, *G. amicalis* LH3 displays the capacity to reduce oil viscosity by 44.7% and degrade oil by 10.4% (w/w) under aerobic conditions. Additionally, this strain exhibits resilience to environmental stressors, thriving in salinity levels of up to 5% NaCl and temperatures ranging from 37 to 40 °C. The supplementation of 0.2% rhamnolipids enhances both the growth and paraffin degradation capabilities of *G. amicalis* LH3, emphasizing its adaptability to enhancement strategies. The multifaceted potential of *G. amicalis* LH3 transcends its degradation capabilities as shown in Fig. 6.2. As a novel long-chain alkane-degrading strain, it holds promise in various applications such as paraffin control, bioremediation of hydrocarbon-contaminated environments, and Microbial Enhanced Oil Recovery (MEOR) initiatives (Hao et al. 2008).

Bacillus subtilis (*Bacillus subtilis* YB7)

Bacillus subtilis, a Gram-positive bacterium, is known for its robustness and adaptability. Recently isolated, it has shown promise in degrading paraffin wax, making it a potential candidate for biodegradation processes, particularly in environments contaminated with paraffin-rich substances. A biosurfactant 'surfactin' is produced by *Bacillus subtilis*. The optimal temperature for biosurfactant production by *Bacillus subtilis* is found to be 50 °C, with maximum yields observed at this temperature (Fig. 6.3). Furthermore, the biosurfactant exhibits stability across a wide pH range, from 8 to 14, indicating its potential applicability in alkaline reservoirs. *Bacillus*

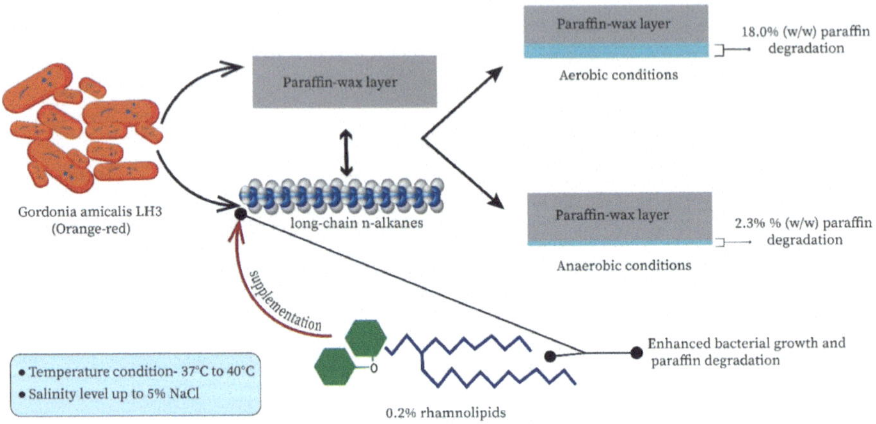

Fig. 6.2 Degradation of paraffin and wax by *Gordonia amicalis* LH3

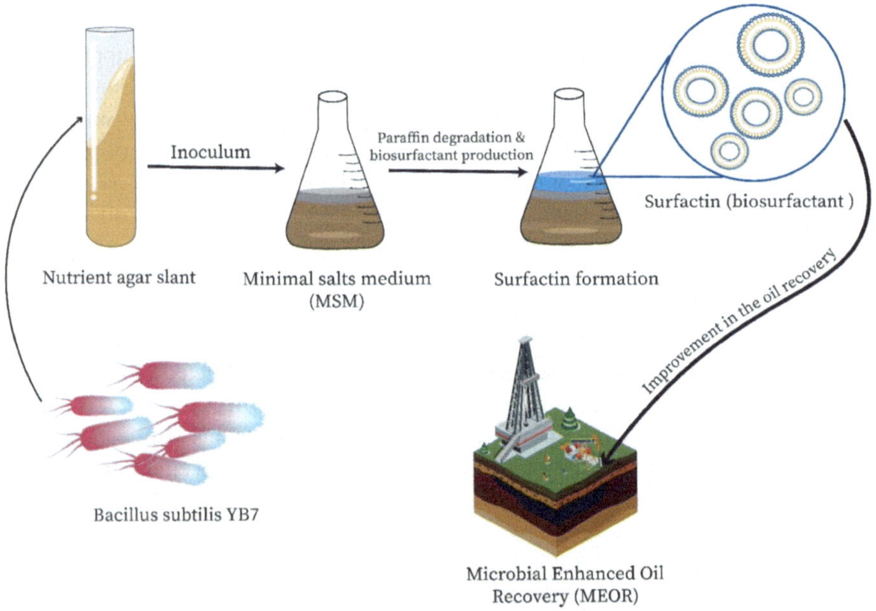

Fig. 6.3 Degradation of paraffin and wax by *Bacillus subtilis* YB7

subtilis demonstrates resilience in environments characterized by extreme conditions such as high alkalinity, temperature, pressure, and salinity, suggesting its suitability for enhanced oil recovery (EOR) and treatment of oil spills (Sakthipriya et al. 2015).

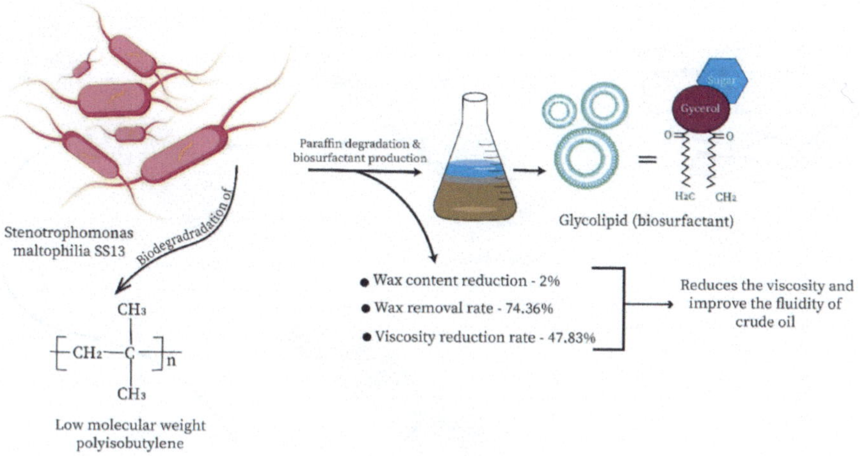

Fig. 6.4 Degradation of paraffin and wax by *Stenotrophomonas maltophilia* SS13

Stenotrophomonas maltophilia (*Stenotrophomonas maltophilia* SS13)

Stenotrophomonas maltophilia is a bacterium recognized for its diverse metabolic capabilities. It has demonstrated the ability to biodegrade low molecular weight polyisobutylene, showcasing its potential role in hydrocarbon degradation processes. The strain produces a biosurfactant (glycolipid) with emulsification properties on liquid paraffin, yielding an emulsification index of 65%. At 37 °C, it demonstrates a wax content reduction of 2%, wax removal rate of 74.36% and a viscosity reduction rate of 47.83% as seen in Fig. 6.4. These findings suggest the strain's effectiveness in wax removal, viscosity reduction, and enhancement of crude oil fluidity, thereby showcasing its practical application value in crude oil exploitation and transportation (Wang et al. 2020).

Rhodococcus erythropolis (*R. erythropolis* DCL14)

Rhodococcus erythropolis is a bacterium known for its hydrocarbon-degrading capabilities. Studies indicate its potential for biodegradation processes, including the breakdown of crude oil, making it a candidate for environmental applications. *R. erythropolis* DCL14 demonstrates the remarkable ability to utilize paraffin wax as its sole carbon and energy source, showcasing its versatility in hydrocarbon degradation. The cells proliferate into a dense biofilm layer on the top of solid substrate, creating scale-like structures that enhance the surface area of the initially smooth paraffin surface while also generating biosurfactants. Previously noted for its capability to degrade C_5 to C_{16} n-alkanes, *R. erythropolis* DCL14 cells exhibit phenotypic adaptations enabling the degradation of solid substrates. Under laboratory conditions, Strain DCL14 displayed efficient degradation, metabolizing 866 mg of paraffin wax

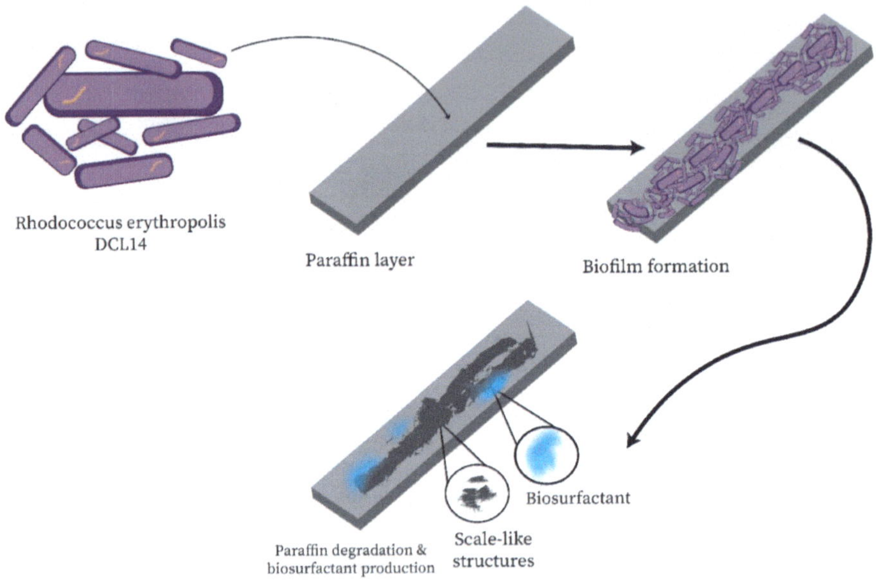

Fig. 6.5 Degradation of paraffin and wax by *Rhodococcus erythropolis* (*R. erythropolis* DCL14)

within 20 mL of mineral medium over a span of 62 h (Fig. 6.5) (Rodrigues and Carvalho 2019). Given its resilience to extreme conditions and its capacity to thrive in fuel oil and various hydrocarbons, it is reasonable to anticipate that *R. erythropolis* DCL14 cells may effectively degrade paraffin wax under real-world scenarios.

6.2 Characterization of Bacterial Strains

6.2.1 Characterization Parameters

Metabolic Versatility: Ability to utilize a broad range of hydrocarbons.

Adaptability: Capacity to thrive in environments with paraffin and wax contaminants.

Enzymatic Capabilities: Presence of enzymes facilitating the breakdown of paraffin and wax.

Hydrocarbon-Degrading Proficiency: Efficient degradation of alkanes and related compounds.

Environmental Adaptation: Adaptation to diverse environments, including oil-producing wells and marine settings.

Potential for Biodegradation: Demonstrated effectiveness in biodegradation processes, including crude oil and heavy oil.

6.2.2 *Pseudomonas sp. Strain PW-1*

This strain demonstrates a moderate metabolic versatility, showcased by its ability to efficiently utilize paraffin wax as a sole carbon source. Its enzymatic activity facilitates the degradation of paraffin wax, leading to a notable degradation rate of 52.85%. Adapted to moderate temperatures, with optimal enzymatic activity observed between 20 and 40 °C, this strain holds potential for biodegradation processes targeting hydrocarbon pollutants, including paraffin wax (Zhang et al. 2016).

6.2.3 *Gordonia amicalis LH3*

Gordonia amicalis LH3 presents intriguing hydrocarbon-degrading capabilities. It efficiently utilizes long-chain n-alkanes as carbon and energy sources and demonstrates significant paraffin degradation under aerobic conditions. The enzymatic activity exhibited by *Gordonia amicalis* LH3 facilitates the breakdown of paraffin, making it an effective agent in paraffin control and bioremediation efforts. Additionally, its resilience to varying environmental conditions, such as salinity levels up to 5% NaCl and temperatures ranging from 37 to 40 °C, further underscores its potential in hydrocarbon-contaminated environments (Hao et al. 2008).

6.2.4 *Bacillus subtilis (Bacillus subtilis YB7)*

Bacillus subtilis shows adaptability to environments contaminated with paraffin-rich substances. Its ability to produce biosurfactants facilitates paraffin degradation, making it suitable for enhanced oil recovery and treatment of oil spills. Moreover, its resilience to extreme environmental conditions suggests potential applications in challenging hydrocarbon-rich environments (Sakthipriya et al. 2015).

6.2.5 *Stenotrophomonas maltophilia (Stenotrophomonas maltophilia SS13)*

Stenotrophomonas maltophilia presents capabilities in hydrocarbon degradation, particularly in the biodegradation of low molecular weight polyisobutylene and paraffin wax. With its ability to produce biosurfactants with emulsification properties, *Stenotrophomonas maltophilia* effectively enhances crude oil fluidity by reducing wax content and viscosity. Its adaptability to various environmental stressors and effectiveness in crude oil exploitation and transportation scenarios position it as a valuable asset in hydrocarbon remediation efforts (Wang et al. 2020).

6.2.6 *Rhodococcus erythropolis DCL14*

Rhodococcus erythropolis DCL14 utilizes paraffin wax as a sole carbon and energy source. Its adaptability to diverse environments, including oil-producing wells, coupled with its efficient degradation of paraffin wax under laboratory conditions, highlights its potential effectiveness in real-world hydrocarbon remediation scenarios. With its capacity to thrive in fuel oil and various hydrocarbons, Rhodococcus erythropolis DCL14 holds promise in addressing paraffin deposition challenges and enhancing oil recovery efforts (Rodrigues and Carvalho 2019). Characterization of bacterial strains from the above studies have been Table 6.1 while optimal temperature and degradation rate (%) of bacterial strains for paraffin and wax degradation is shown in Table 6.2.

6.3 Conclusion

The exploration of paraffin and wax-degrading bacterial strains represents a significant advancement in environmental biotechnology, offering promising solutions for hydrocarbon remediation. The characterization of selected bacterial strains, such as *Pseudomonas* sp. PW-1, *Gordonia amicalis* LH3, *Bacillus subtilis*, *Stenotrophomonas maltophilia*, and *Rhodococcus erythropolis* DCL14, highlights their diverse metabolic capabilities and enzymatic pathways tailored for the degradation of complex hydrocarbon structures. These bacterial strains demonstrate remarkable potential in addressing challenges associated with paraffin and wax contamination in various ecological settings. Their ability to efficiently degrade paraffin and wax compounds not only contributes to the reduction of environmental pollution but also promotes ecological balance and sustainability. Moreover, the adaptability of these bacterial strains to diverse environmental conditions, including oil-producing wells and marine settings, underscores their versatility and applicability in real-world scenarios. Their resilience to environmental stressors and their effectiveness

Table 6.1 Characterization of bacterial strains from respective studies

Name of strain	Sites of microorganism isolation	Characteristics of Microorganisms	Inoculum	Main results	References
Pseudomonas sp. strain PW-1	Daqing oil-contaminated sites in China	Gram-negative rod shaped	Oil field	Strain effectively removes paraffin wax that causes deposition in oil wells	Zhang et al. (2016)
Gordonia amicalis LH3	Jidong Oilfield Co. Ltd., in Tangshan, China	Circular, smooth, orange-red, convex elevation, short rod-shaped	Oil-contaminated water samples	*G. amicalis* LH3 evidenced the highest values for both oil viscosity reduction and paraffin degradation	Hao et al. (2008)
Bacillus subtilis YB7	Dump site, Chennai, 140 India	Gram positive and rod shaped	Dump site	*Bacillus subtilis* can degrade paraffin in environment where extreme conditions of alkalinity, temperature, pressure and salt exist	Sakthipriya et al. (2015), Arutchelvi et al. (2009)
Stenotrophomonas maltophilia SS13	Liaohe Oilfield, China	Gram-negative	Soil sample heavily polluted by the crude oil	The strain can effectively remove the wax, reduce the viscosity of crude oil, and improve the fluidity of crude oil	Wang et al. (2020)
Rhodococcus erythropolis (*R. erythropolis* DCL14)	Reeuwijk, The Netherlands	Gram-positive, partially acid-fast, coccoid to rod-shaped	Sample deposited and maintained at the iBB-Institute for Bioengineering and Biosciences, Instituto Superior Técnico, Portugal	*R. erythropolis* DCL14 cells can degrade solid substrate by making phenotypic adaptations. Can degrade paraffin	Rodrigues and Carvalho (2019), Bagdure et al. (2012)

Table 6.2 Optimal temperature and degradation rate (%) of bacterial strains for paraffin and wax degradation

Bacterial strain	Optimal temperature (°C)	Degradation rate (%)
Pseudomonas sp. strain PW-1	20–40	52.85%
Gordonia amicalis LH3	37–40	18.0% (w/w)—under aerobic conditions 2.3% (w/w)—under anaerobic conditions
Bacillus subtilis YB7	50	–
Stenotrophomonas maltophilia SS13	37	74.36%
Rhodococcus erythropolis (*R. erythropolis* DCL14)	55	Degradation of 866 mg of paraffin wax within 20 mL of mineral medium over a span of 62 h

in enhancing crude oil recovery further emphasize their value in hydrocarbon remediation efforts. Looking ahead, future research should focus on several key areas to harness the full potential of paraffin and wax-degrading bacterial strains. First, there is a need for in-depth exploration of the metabolic pathways and enzymatic mechanisms involved in hydrocarbon degradation to optimize bioremediation strategies. Understanding the ecological implications of bacterial-mediated degradation processes is also crucial for assessing the long-term sustainability of remediation efforts. Furthermore, the development of innovative biotechnological approaches, such as genetic engineering and synthetic biology, holds promise for enhancing the efficiency and specificity of hydrocarbon remediation processes. By leveraging advances in biotechnology, researchers can design tailored solutions for targeted degradation of paraffin and wax contaminants, minimizing environmental impact and promoting ecosystem health. In conclusion, the continued exploration and utilization of paraffin and wax-degrading bacterial strains offer immense potential for mitigating hydrocarbon pollution and restoring contaminated environments. Through interdisciplinary collaboration and innovative research endeavors, we can unlock new opportunities for sustainable environmental stewardship and conservation.

References:

Arutchelvi J et al (2009) Production and characterization of biosurfactant from Bacillus subtilis YB7. J Appl Sci 9

Azubuike CC, Chikere CB, Okpokwasili GC (2020) Bioremediation: an eco-friendly sustainable technology for environmental management. In: Saxena G, Bharagava RN (eds) Bioremediation of industrial waste for environmental safety: volume i: industrial waste and its management. Springer Singapore, Singapore, pp 19–39

Bagdure SR et al (2012) Rhodococcus erythropolis encephalitis in patient receiving rituximab. Emerg Infect Dis 18(8):1377–1379

Feild TS et al (2005) Hydathodal leaf teeth of Chloranthus japonicus (Chloranthaceae) prevent guttation-induced flooding of the mesophyll. Plant, Cell Environ 28(9):1179–1190

Griffin MR (2014) Biocontrol and bioremediation: two areas of endophytic research which hold great promise. In: Verma VC, Gange AC (eds) Advances in endophytic research. Springer India, New Delhi, pp 257–282

Hadley NF (1989) Lipid water barriers in biological systems. Prog Lipid Res 28(1):1–33

Hao D-H et al (2008) Isolation, identification, and performance studies of a novel paraffin-degrading bacterium of Gordonia amicalis LH3. Biotechnol Bioprocess Eng 13(1):61–68

Ławniczak Ł et al (2020) Microbial degradation of hydrocarbons—basic principles for bioremediation: a review. Molecules 25(4):856

Mansoori GA, Barnes HL, Webster GM (2003) Petroleum waxes. Fuels and lubricants handbook 19:525–558

Misra S, Baruah S, Singh K (1995) Paraffin problems in crude oil production and transportation: a review. SPE Prod Facil 10(01):50–54

Mohanan N et al (2020) Microbial and enzymatic degradation of synthetic plastics. Front Microbiol 11:580709

Ossai IC et al (2020) Remediation of soil and water contaminated with petroleum hydrocarbon: a review. Environ Technol Innov 17:100526

Rodrigues CJ, de Carvalho CC (2019) Phenotypic adaptations help Rhodococcus erythropolis cells during the degradation of paraffin wax. Biotechnol J 14(8):1800598

Sakthipriya N, Doble M, Sangwai JS (2015) Action of biosurfactant producing thermophilic Bacillus subtilis on waxy crude oil and long chain paraffins. Int Biodeterior Biodegrad 105:168–177

Silva IA et al (2022) Oil spills: impacts and perspectives of treatment technologies with focus on the use of green surfactants. Environ Monit Assess 194(3):143

Theyab M (2020) A review of wax mitigation methods through hydrocarbon production. J Pet Environ Biotechnol 9:412

Van Hamme Jonathan D, Singh A, Ward Owen P (2003) Recent advances in petroleum microbiology. Microbiol Mol Biol Rev 67(4):503–549

Wang W et al (2020) Determination of wax removal and viscosity reduction in crude oil treated by dominant bacteria. Pet Sci Technol 38(7):627–634

Wentzel A et al (2007) Bacterial metabolism of long-chain n-alkanes. Appl Microbiol Biotechnol 76(6):1209–1221

Zhang YL, Liu Z, Liu T (2016) Isolation and characterization of a novel paraffin wax-degrading bacterium, Pseudomonas sp. strain PW-1, from petroleum-contaminated sites. Genet Mol Res 15(2)

Chapter 7
Isolated Recyclable Environmental Friendly Surfactants for Potential Degradation of Waxes and Paraffins in Oil and Gas Industries

Gourav Patil, Prasenjit Chakraborty, and Shirsendu Banerjee

Abstract The oil and gas industry faces significant challenges posed by the deposition of waxes and paraffins in crude oil and natural gas streams, necessitating effective strategies for their degradation. Conventional methods for wax management often involve chemical inhibitors or mechanical interventions, which may not align with sustainability objectives. In response, the exploration of environmentally friendly surfactants capable of efficiently degrading waxes and paraffins has gained traction. Biosurfactants, derived from renewable biological sources, emerge as promising alternatives to petroleum-based surfactants due to their biodegradability, low toxicity, and renewable nature. The isolation and characterization of biosurfactants involve various methods, including solvent extraction and purification techniques, to obtain high-purity products suitable for industrial applications. These biosurfactants exhibit physicochemical properties such as surface tension reduction and emulsification, essential for disrupting wax deposition and facilitating their removal. This review explores the role of isolated recyclable environmental-friendly surfactants in addressing wax and paraffin deposition challenges in the oil and gas industry.

G. Patil
School of Biotechnology, Kalinga Institute of Industrial Technology, Bhubaneswar, India

P. Chakraborty
Department of Chemistry, Burdwan University, Burdwan, India

S. Banerjee (✉)
School of Chemical Engineering, Kalinga Institute of Industrial Technology, Bhubaneswar, India
e-mail: shirsendu.banerjee@kiitbiotech.ac.in

S. Banerjee et al. (eds.), *Environmental Friendly Green Technologies for Improvement of Heavy Crude Oil Flow Assurance*, Environmental Science and Engineering,
https://doi.org/10.1007/978-3-031-86701-9_7

7.1 Introduction

7.1.1 Context of Environmentally Friendly Surfactants in the Oil and Gas Industry

Surfactants play a crucial role in various industrial processes, particularly within the oil and gas sector, where they are employed for tasks such as drilling fluids and enhanced oil recovery techniques (Quintero 2002). Their amphiphilic properties enable them to reduce interfacial tension, facilitating processes such as emulsification and dispersion of hydrophobic and hydrophilic components. However, conventional surfactants used in these applications often raise environmental concerns due to their persistence, toxicity, and potential for bioaccumulation (Nagtode et al. 2023; Eras-Muñoz et al. 2022). The pursuit of environmentally friendly alternatives has gained momentum in response to these challenges. Environmentally friendly surfactants are those designed to minimize ecological impacts while maintaining performance standards. These may include surfactants derived from renewable sources, those with biodegradable properties, and those exhibiting low toxicity levels (Nagtode et al. 2023). Within the oil and gas industry, the demand for environmentally friendly surfactants is fueled by escalating environmental regulations, corporate sustainability agendas, and societal pressure to reduce industrial environmental footprints (Gautam et al. 2023). Furthermore, the adverse effects of conventional surfactants on aquatic ecosystems, soil quality, and human health underscore the urgent need to transition towards more sustainable alternatives.

7.1.2 Why Surfactants Capable of Efficiently Degrading Waxes and Paraffins Are Needed

Waxes and paraffins are ubiquitous components in crude oil and natural gas streams, presenting significant challenges throughout production, transportation, and processing (Robinson and Hsu 2017). Their deposition on equipment surfaces can lead to flow assurance issues such as pipeline blockages and decreased efficiency of production facilities. Traditional methods for mitigating wax and paraffin deposition often involve chemical inhibitors or mechanical interventions, which may not be environmentally sustainable in the long term. The development of surfactants capable of effectively degrading waxes and paraffins offers a promising solution to these challenges. By leveraging the inherent surface-active properties of surfactants, it becomes possible to disrupt the molecular interactions that contribute to wax deposition and facilitate their dispersion or solubilization in the surrounding medium. This not only helps prevent the formation of wax deposits but also enables the removal of existing deposits through enhanced cleaning processes. The search for surfactants that can degrade wax and paraffin is driven by the need to make operations more efficient, cut down on maintenance expenses, and lessen the environmental harm caused by conventional wax management techniques (Theyab 2020). Furthermore, deploying such surfactants aligns with broader sustainability objectives within the oil and gas industry, promoting resource conservation and mitigating ecological harm.

7.2 Biosurfactants: A Green Alternative

7.2.1 *Advantages of Biosurfactants Over Petroleum-Based Surfactants*

Biosurfactants have emerged as a promising green alternative to petroleum-based surfactants due to several advantageous properties. Unlike their synthetic counterparts, biosurfactants are produced from renewable biological sources, offering a sustainable solution to surfactant demand (Banat et al. 2010). One of the key advantages of biosurfactants is their diverse chemical structures and functionalities, which result in superior performance in various applications, including the oil and gas industry (Cássia et al. 2014). Additionally, biosurfactants exhibit high surface activity and stability over a wide range of environmental conditions, making them suitable for use in harsh industrial settings (Shaikhah et al. 2024). Furthermore, biosurfactants often demonstrate enhanced biodegradability compared to petroleum-based surfactants, mitigating concerns related to environmental persistence and toxicity (Marchant and Banat 2012). This biodegradability not only reduces the environmental footprint associated with surfactant use but also contributes to ecosystem preservation and remediation efforts (Nagtode et al. 2023). Moreover, biosurfactants are compatible with existing wastewater treatment processes, facilitating their integration into industrial operations without significant infrastructure modifications (Satpute et al. 2010).

7.2.2 *Biodegradability, Low Toxicity, and Renewable Sources*

Biosurfactants are characterized by their biodegradability, low toxicity, and renewable sources, distinguishing them from petroleum-based surfactants. Biodegradability ensures that biosurfactants can be broken down by microbial action into harmless byproducts, reducing their environmental impact and potential for bioaccumulation (Bjerk et al. 2021). This property is particularly advantageous in applications where surfactants come into contact with soil, water bodies, or living organisms, as it minimizes ecological disruption and contamination (Gudiña et al. 2016). Additionally, biosurfactants exhibit low toxicity levels, making them safer for human health and environmental ecosystems compared to their synthetic counterparts (Johnson et al. 2021). This low toxicity profile allows for broader use in consumer products, pharmaceuticals, and agricultural applications without compromising safety standards (Banat et al. 2010). Moreover, biosurfactants derived from renewable biological sources such as microbes, plants, and agricultural residues offer a sustainable alternative to finite petroleum resources (Marchant and Banat 2012). The renewable nature of these sources ensures a continuous and environmentally friendly supply chain for biosurfactant production, reducing dependence on fossil fuels and contributing to a circular economy (Shaikhah et al. 2024).

7.3 Isolation and Characterization of Biosurfactants

7.3.1 Methods for Isolating and Purifying Biosurfactants

The isolation and purification of biosurfactants typically involve a series of extraction, separation, and purification steps to obtain a high-purity product suitable for characterization and application. One common method for biosurfactant isolation is solvent extraction, where the biosurfactant-producing microorganisms are cultured in a suitable growth medium, and the biosurfactants are extracted from the culture broth using organic solvents such as chloroform, ethyl acetate, or butanol (Marchant and Banat 2012; Banat et al. 2010). After extraction, the crude biosurfactant extract undergoes purification to remove impurities and concentrate the biosurfactant compound. Techniques such as column chromatography, ultrafiltration, and precipitation with salts or acids are commonly employed for this purpose (Satpute et al. 2010). Column chromatography separates the biosurfactant from other components based on differences in their chemical properties, such as polarity or molecular weight, while ultrafiltration selectively removes smaller molecules and impurities through a semipermeable membrane. Furthermore, advanced techniques such as high-performance liquid chromatography (HPLC) and mass spectrometry (MS) can be utilized for the precise separation and identification of individual biosurfactant molecules within the crude extract (Gudiña et al. 2016). These techniques enable researchers to characterize the chemical structure, composition, and purity of the isolated biosurfactants with high accuracy and sensitivity (Fig. 7.1).

7.3.2 Physicochemical Properties (Surface Tension Reduction, Emulsification)

Biosurfactants exhibit various physicochemical properties that contribute to their surface-active behavior and emulsification capabilities. One of the primary functions of biosurfactants is the reduction of surface tension at the interface between two immiscible phases, such as oil and water (Banat et al. 2010). This surface tension reduction facilitates the dispersion and solubilization of hydrophobic compounds in

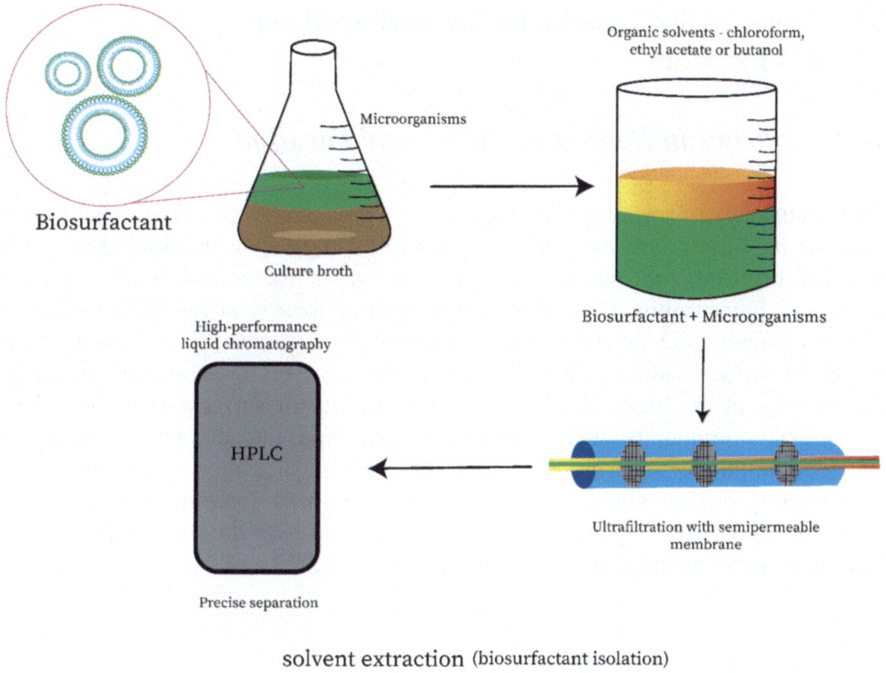

Fig. 7.1 Biosurfactant isolation using solvent extraction

aqueous solutions, leading to increased bioavailability and enhanced bioremediation of pollutants. The effectiveness of biosurfactants in reducing surface tension is often quantified using methods such as the pendant drop technique or the Wilhelmy plate method, which measures the equilibrium surface tension of the solution (Wojciechowski et al. 2023). Biosurfactants can achieve significant reductions in surface tension, typically below 40 mN/m, depending on their chemical structure and concentration (Santos et al. 2016). In addition to surface tension reduction, biosurfactants also possess emulsification properties, enabling them to stabilize emulsions and facilitate the formation of oil-in-water or water-in-oil emulsions (Santos et al. 2016). This emulsification capability is crucial in various industrial applications, including food processing, cosmetics, and pharmaceuticals, where stable emulsions are desired for product formulation and delivery. Overall, the physicochemical properties of biosurfactants, including their surface tension reduction and emulsification capabilities, make them versatile and environmentally friendly alternatives to synthetic surfactants in numerous industrial and environmental applications.

7.4 Potential Surfactants for Degrading Waxes and Paraffins

7.4.1 *Efficacy in Degrading Waxes and Paraffins*

The effectiveness of surfactants in degrading waxes and paraffins is a crucial aspect of addressing flow assurance challenges in the oil and gas industry. Surfactants with the capability to disrupt the molecular interactions of waxes and paraffins play a pivotal role in preventing deposition and facilitating their removal. The efficacy of these surfactants is evaluated based on their ability to reduce wax deposition on equipment surfaces, improve fluid flow, and enhance overall operational efficiency (Eras-Muñoz et al. 2022). Additionally, the capacity of surfactants to solubilize hydrophobic compounds, such as waxes and paraffins, into the aqueous phase is a key determinant of their efficacy in degrading these substances. Surfactants that exhibit strong surface activity and emulsification properties demonstrate enhanced performance in dispersing and solubilizing waxes and paraffin, thereby preventing their agglomeration and deposition (Al-Shboul et al. 2023).

7.4.2 *Notable Biosurfactants*

7.4.2.1 Rhamnolipids

Rhamnolipids are among the most well-known biosurfactants with proven efficacy in degrading hydrocarbons, including waxes and paraffins. Produced by various Pseudomonas species, rhamnolipids exhibit exceptional surface activity and emulsification properties, allowing them to effectively solubilize and disperse hydrophobic compounds (Mulligan 2005). Recent studies showed successful production of biosurfactants from Pseudomonas aeruginosa and Pseudomonas fluorescens, utilizing n-hexadecane and n-eicosane as carbon sources. The organisms yielded 9.8 and 8.2 g L^{-1} of rhamnolipid containing C8 and C10 fatty acids, respectively. Remarkably, both microorganisms demonstrated the ability to degrade 85–90% of the waxes within a single day. Furthermore, the biosurfactant exhibited impressive stability, withstanding temperatures up to 100 °C, pressures of 8 MPa, salinity levels of 20% (w/v), and alkaline pH conditions. The research underscores the potential of these microorganisms in biosurfactant production and highlights the robust properties of the resulting product. Their ability to reduce interfacial tension and enhance oil recovery makes them promising candidates for addressing wax and paraffin deposition issues in the oil and gas industry (Sakthipriya et al. 2015). Similarly, the bacillus amyloliquefaciens 6-2c bacterial strain has demonstrated remarkable potential. Notably, this bacterial strain has exhibited efficiency in enhancing the solubility of paraffin in n-hexane, a crucial attribute for facilitating the removal of paraffin deposits in oil reservoirs. The study has revealed the notable impact of strain 6-2c

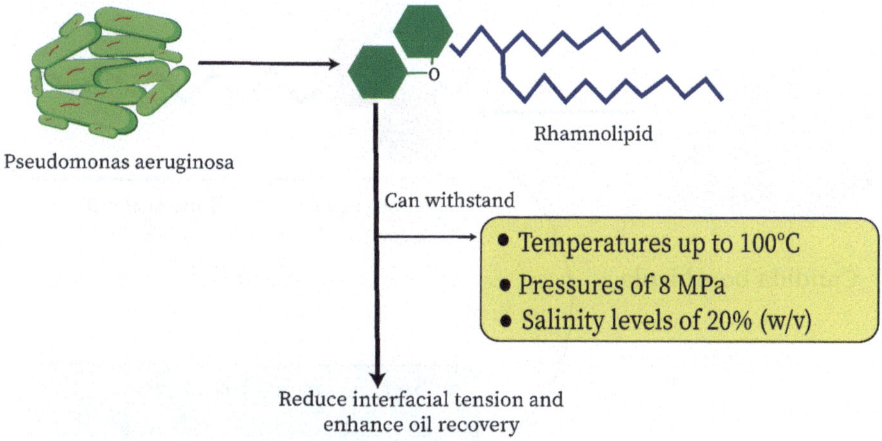

Fig. 7.2 Structural depiction of Rhamnolipids and working condition parameters

on paraffin removal efficiency, showcasing a broad range of efficacy, spanning from 50.82% to an impressive 95.17%. This variability suggests adaptability to different environmental conditions, indicative of its versatility in diverse oil recovery settings. Particularly strain 6-2c cells show the ability to adhere to the surfaces of steel pipes, effectively inhibiting paraffin coalescence. This phenomenon demonstrates the multifaceted role of this biosurfactant-producing bacterial strain in mitigating paraffin-related challenges encountered in oil production processes (Zhang et al. 2018) (Fig. 7.2).

7.4.2.2 Sophorolipids: Derived from Yeast, Effective in Emulsifying and Degrading Hydrocarbons

Sophorolipids, produced by yeast strains such as Candida bombicola, are versatile biosurfactants known for their emulsification and degradation capabilities. These surfactants exhibit strong surface activity and are capable of forming stable emulsions with hydrophobic compounds, including waxes and paraffins. Sophorolipids have demonstrated promising results in enhancing hydrocarbon recovery and reducing wax deposition in oil production processes (Elshafie et al. 2015). Sophorolipids are amphiphilic molecules that consist of a hydrophilic carbohydrate head group linked to one or two hydrophobic fatty acid chains (Cho et al. 2022). The unique structure of sophorolipids imparts them with exceptional surface activity and emulsification properties, making them highly effective in interacting with hydrophobic compounds such as waxes and paraffins (Totsingan et al. 2021). The production of sophorolipids from renewable feedstocks, including vegetable oils, fatty acids, and hydrocarbons, offers a sustainable and environmentally friendly alternative to petroleum-based surfactants (Wongsirichot et al. 2021). The biodegradability and

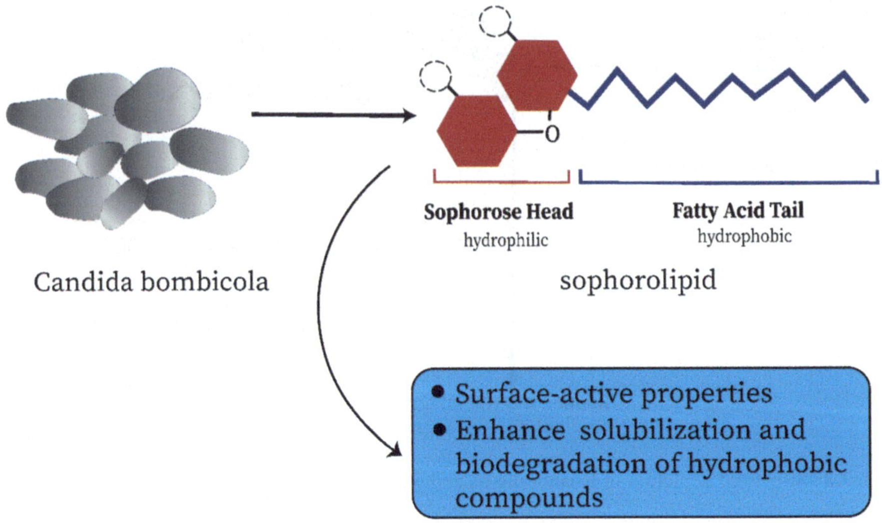

Fig. 7.3 Structure and functions of sophorolipids

low toxicity of sophorolipids further contribute to their eco-friendly profile, ensuring minimal environmental impact during their use and disposal. In addition to their surface-active properties, sophorolipids possess remarkable emulsification capabilities, allowing them to form stable oil-in-water and water-in-oil emulsions. This ability makes them suitable for applications in diverse industries, including bioremediation, food processing, and cosmetics (Pal et al. 2023). Studies have demonstrated the efficacy of sophorolipids in enhancing the solubilization and biodegradation of hydrophobic compounds, including crude oil components, in contaminated environments. Sophorolipids act by lowering interfacial tension, promoting the dispersion of hydrophobic substances, and facilitating their uptake and metabolism by indigenous microorganisms (Miao 2024). Moreover, the versatility of sophorolipids extends to their ability to function under a wide range of environmental conditions, including variations in temperature, pH, and salinity. This robustness makes sophorolipids suitable for use in diverse industrial settings, including oilfield applications where extreme conditions are present (Oliveira et al. 2015) (Fig. 7.3).

7.4.2.3 Trehalolipids: Produced by Rhodococcus and Corynebacterium, Exhibit Strong Surface Activity

Trehalolipids are glycolipid biosurfactants produced by various microbial species, including Rhodococcus and Corynebacterium. These surfactants possess significant surface activity and emulsification properties, enabling them to effectively interact with hydrophobic compounds (Andreolli et al. 2023). Trehalolipids have been shown to enhance the solubilization and degradation of waxes and paraffins, making them promising candidates for applications in the oil and gas industry. Trehalolipids have been found to exhibit strong surface activity and emulsification properties, enabling them to effectively interact with hydrophobic compounds such as waxes and paraffin (Janek et al. 2018; Suzuki et al. 1969). Their ability to reduce interfacial tension and enhance solubilization makes them promising candidates for applications in the oil and gas industry, particularly in mitigating wax deposition issues and improving oil recovery efficiency. Their compatibility with microbial degradation processes ensures minimal environmental impact and facilitates their use in bioremediation and wastewater treatment applications. Research studies have demonstrated the efficacy of trehalolipids in enhancing oil recovery and reducing wax deposition in pipelines and equipment surfaces (Filho et al. 2023). Their strong surface activity and biodegradable nature make them promising candidates for sustainable solutions in the oil and gas industry, aligning with the industry's efforts towards environmental stewardship and resource efficiency.

7.4.2.4 Lipopeptides: Produced by Bacillus Species, Show Promise in Hydrocarbon Degradation

Lipopeptides are amphiphilic molecules synthesized by Bacillus species, known for their ability to solubilize and degrade hydrophobic compounds. These surfactants exhibit strong surface activity and emulsification properties, making them effective in dispersing and degrading waxes and paraffin (Jacques and Surfactin and Other Lipopeptides from Bacillus spp. 2010; Feng et al. 2019). Lipopeptides have demonstrated potential in enhancing oil recovery and mitigating wax deposition issues in oil and gas production processes. Lipopeptides have garnered significant attention for their exceptional surface activity, emulsification capacity, and ability to solubilize hydrophobic (Raaijmakers et al. 2010). Their unique molecular structure allows them to adsorb at oil–water interfaces, reducing interfacial tension and stabilizing emulsions containing hydrophobic compounds. This property facilitates the dispersion and degradation of waxes and paraffins, preventing their deposition and buildup on equipment surfaces in oil production and transportation systems. lipopeptides exhibit strong bioactivity against various microorganisms, including those involved in hydrocarbon degradation processes (Banat et al. 2000). They can enhance the growth and activity of indigenous hydrocarbon-degrading bacteria, promoting the biodegradation of waxes and paraffin in contaminated environments. Additionally, lipopeptides have been shown to inhibit the growth of pathogenic microorganisms, contributing to the preservation of environmental and human health. Research studies have highlighted the potential of lipopeptides in enhancing oil recovery efficiency and mitigating wax deposition issues in pipelines and production facilities (Zhang et al. 2016). Their biocompatibility, biodegradability, and low toxicity make them attractive options for sustainable solutions in the oil and gas industry, aligning with the industry's growing emphasis on environmental stewardship and resource efficiency (Table 7.1).

Table 7.1 Comparison of Microbial Surfactants for Wax and Paraffin Degradation in the Oil and Gas Industry

Surfactant	Source	Mode of action	Stability	Biodegradability	References
Rhamnolipids	Pseudomonas spp.	Emulsification and Bioemulsification	Moderate	High	Abdel-Mawgoud et al. (2010), Das et al. (2014)
Sophorolipids	Candida spp.	Solubilization and dispersal	High	High	Bogaert et al. (2007), Akubude and Mba (2021)
Trehalolipids	Mycobacteria spp.	Disruption of microbial membranes	High	High	Nitschke and Costa (2007)
Lipopeptides	Bacillus spp.	Disruption of cell walls	High	High	Raaijmakers et al. (2010), Saranraij et al. (2022)

7.5 Concluding Remarks and Future Perspectives

The development and application of environmentally friendly surfactants present a promising pathway for addressing wax and paraffin deposition challenges in the oil and gas industry. Biosurfactants, with their biodegradability, low toxicity, and renewable nature, offer sustainable alternatives to conventional surfactants. The efficacy of biosurfactants such as rhamnolipids, sophorolipids, trehalolipids, and lipopeptides in degrading waxes and paraffins underscore their potential for enhancing operational efficiency while minimizing environmental impact. Moving forward, further research is essential to optimize the production, characterization, and application of environmentally friendly surfactants. Investigating novel biosurfactant-producing microorganisms, refining extraction and purification techniques, and exploring synergistic formulations could enhance surfactant performance and scalability. Additionally, studies focusing on the long-term environmental impact and cost-effectiveness of biosurfactants compared to traditional methods will provide valuable insights for industry stakeholders. Collaborative efforts between academia, industry, and regulatory bodies are crucial for driving innovation and adoption of sustainable wax management practices in the oil and gas sector, ultimately contributing to a more environmentally conscious and resilient industry landscape.

References

Abdel-Mawgoud AM, Lépine F, Déziel E (2010) Rhamnolipids: diversity of structures, microbial origins and roles. Appl Microbiol Biotechnol 86(5):1323–1336

Akubude VC, Mba BA (2021) Chapter 4—Application of biosurfactants in algae cultivation systems. In: Green sustainable process for chemical and environmental engineering and science, Inamuddin, C.O. Adetunji, and A.M. Asiri, Editors. 2021, Elsevier, pp 97–108

Al-Shboul T, Sagala F, Nassar NN (2023) Role of surfactants, polymers, nanoparticles, and its combination in inhibition of wax deposition and precipitation: a review. Adv Coll Interface Sci 315:102904

Andreolli M et al (2023) Characterization of trehalolipid biosurfactant produced by the novel marine strain Rhodococcus sp. SP1d and its potential for environmental applications. Microb Cell Fact 22(1):126

Banat IM et al (2010) Microbial biosurfactants production, applications and future potential. Appl Microbiol Biotechnol 87(2):427–444

Banat IM, Makkar RS, Cameotra SS (2000) Potential commercial applications of microbial surfactants. Appl Microbiol Biotechnol 53(5):495–508

Bjerk TR et al (2021) Biosurfactants: properties and applications in drug delivery, biotechnology and ecotoxicology. Bioengineering (Basel) 8(8)

Cho WY et al (2022) Sophorolipids-bio-based antimicrobial formulating agents for applications in food and health. Molecules 27(17)

de Cássia FSSR et al (2014) Applications of biosurfactants in the petroleum industry and the remediation of oil spills. Int J Mol Sci 15(7):12523–12542

Das P, Yang XP, Ma LZ (2014) Analysis of biosurfactants from industrially viable Pseudomonas strain isolated from crude oil suggests how rhamnolipids congeners affect emulsification property and antimicrobial activity. Front Microbiol 5:696

Elshafie AE et al (2015) Sophorolipids production by candida bombicola ATCC 22214 and its potential application in microbial enhanced oil recovery. Front Microbiol 6:1324

Eras-Muñoz E et al (2022) Microbial biosurfactants: a review of recent environmental applications. Bioengineered 13(5):12365–12391

Feng JQ et al (2019) Characterization of biosurfactant lipopeptide and its performance evaluation for oil-spill remediation. RSC Adv 9(17):9629–9632

Filho A et al (2023) Biosurfactants as multifunctional remediation agents of environmental pollutants generated by the petroleum industry. Energies 16:1209

Gautam K et al (2023) Oily waste to biosurfactant: A path towards carbon neutrality and environmental sustainability. Environ Technol Innov 30:103095

Gudiña EJ, Teixeira JA, Rodrigues LR (2016) Biosurfactants produced by marine microorganisms with therapeutic applications. Mar Drugs 14(2)

Jacques P (2010) Surfactin and other lipopeptides from bacillus, pp 57–91

Janek T et al (2018) Trehalose lipid biosurfactant reduces adhesion of microbial pathogens to polystyrene and silicone surfaces: an experimental and computational approach. Front Microbiol 9:2441

Johnson P et al (2021) Effect of synthetic surfactants on the environment and the potential for substitution by biosurfactants. Adv Coll Interface Sci 288:102340

Marchant R, Banat IM (2012) Microbial biosurfactants: challenges and opportunities for future exploitation. Trends Biotechnol 30(11):558–565

Miao Y et al (2024) Sustainable biosurfactant production from secondary feedstock—recent advances, process optimization and perspectives. Front Chem, 12

Mulligan CN (2005) Environmental applications for biosurfactants. Environ Pollut 133(2):183–198

Nagtode VS et al (2023) Green surfactants (Biosurfactants): a petroleum-free substitute for sustainability–comparison, applications, market, and future prospects. ACS Omega 8(13):11674–11699

Nitschke M, Costa SGVAO (2007) Biosurfactants in food industry. Trends Food Sci Technol 18(5):252–259
Oliveira M et al (2015) Review: sophorolipids a promising biosurfactant and it's applications. Int J Adv Biotechnolo Res (IJBR) 6:161–174
Pal S et al (2023) Sophorolipids: a comprehensive review on properties and applications. Adv Coll Interface Sci 313:102856
Quintero L (2002) An overview of surfactant applications in drilling fluids for the petroleum industry. J Dispersion Sci Technol 23(1–3):393–404
Raaijmakers JM et al (2010) Natural functions of lipopeptides from Bacillus and Pseudomonas: more than surfactants and antibiotics. FEMS Microbiol Rev 34(6):1037–1062
Robinson PR, Hsu CS (2017) Introduction to petroleum technology. Springer handbook of petroleum technology, pp 1–83
Sakthipriya N, Doble M, Sangwai JS (2015) Biosurfactant from Pseudomonas species with waxes as carbon source—their production, modeling and properties. J Ind Eng Chem 31:100–111
Santos DK et al (2016) Biosurfactants: multifunctional biomolecules of the 21st century. Int J Mol Sci 17(3):401
Saranraij P et al (2022) Microbial fermentation technology for biosurfactants production, pp 25–43
Satpute SK et al (2010) Biosurfactants, bioemulsifiers and exopolysaccharides from marine microorganisms. Biotechnol Adv 28(4):436–450
Shaikhah D et al (2024) New trends in biosurfactants: from renewable origin to green enhanced oil recovery applications. Molecules 29(2)
Suzuki T et al (1969) Trehalose lipid and α-branched-β-hydroxy fatty acid formed by bacteria grown on n-alkanes. Agric Biol Chem 33(11):1619–1627
Theyab M (2020) A review of wax mitigation methods through hydrocarbon production. J Pet Environ Biotechnol 9:412
Totsingan F, Liu F, Gross RA (2021) Structure-activity relationship assessment of sophorolipid ester derivatives against model bacteria strains. Molecules 26(10)
Van Bogaert IN et al (2007) Microbial production and application of sophorolipids. Appl Microbiol Biotechnol 76(1):23–34
Wojciechowski K, Borucka K, Mierzejewska J (2023) Are all yeast biosurfactants really capable of lowering surface tension below 30 mN/m ? Colloids Surf, B 230:113503
Wongsirichot P, Ingham B, Winterburn J (2021) A review of sophorolipid production from alternative feedstocks for the development of a localized selection strategy. J Clean Prod 319:128727
Zhang J et al (2016) Production of lipopeptide biosurfactants by Bacillus atrophaeus 5–2a and their potential use in microbial enhanced oil recovery. Microb Cell Fact 15(1):168
Zhang J et al (2018) Prevention and mitigation of paraffin deposition by biosurfactant-producing and paraffin-degrading Bacillus amyloliquefaciens strain 6–2c. Chem Eng J 335:510–519

Chapter 8
Applications of Recyclable Environmental Friendly Fungi in the Degradation of Paraffin, Waxes and Asphaltenes in Oil and Gas Industries

Ramandeep Saini and Ayush Madan

Abstract The accumulation of paraffin, waxes, and asphaltenes in the oil and gas industries presents significant challenges, impacting production and transportation processes and resulting in reduced efficiency and increased operational costs. Conventional methods for their removal and degradation often rely on harsh chemicals or costly physical techniques, which are both environmentally detrimental and economically unsustainable. In recent years, there has been a growing interest in exploring alternative, eco-friendly approaches for addressing these hydrocarbon compounds. This chapter provides an overview of various types of fungi that have been investigated for the degradation of paraffin, waxes, and asphaltenes, along with the enzymatic mechanisms involved in the degradation process. Emphasis has been given to the challenges associated with fungal degradation and the strategies employed to optimize their performance in industrial settings. Additionally, the potential synergistic effects of combining fungi with other biological agents or chemical treatments are explored.

Keywords Paraffin · Waxes · Asphaltenes · Recyclable fungi · Eco-friendly degradation · Enzymatic mechanisms

---

R. Saini
Chandigarh College of Technology, Chandigarh Group of Colleges, Greater Mohali, Punjab, India

A. Madan (✉)
Department of Biotechnology, School of Research & Technology, People's University, Bhopal, Madhya Pradesh, India
e-mail: madanayushmadan@gmail.com

Institute of Tropical Aquaculture and Fisheries (AKUATROP), Universiti Malaysia Terengganu, Terengganu, Malaysia

S. Banerjee et al. (eds.), *Environmental Friendly Green Technologies for Improvement of Heavy Crude Oil Flow Assurance*, Environmental Science and Engineering,
https://doi.org/10.1007/978-3-031-86701-9_8

8.1 Introduction

The oil and gas industry, integral to global energy demands, has long grappled with environmental challenges arising from the extraction, processing, and transportation of hydrocarbons. Among these challenges, the persistence of paraffin, waxes, and asphaltenes in various stages of production poses a significant threat to ecosystems and human health. Traditional methods of hydrocarbon remediation often involve harsh chemicals and energy-intensive processes, contributing to further environmental degradation. In light of these concerns, there is a growing imperative to explore sustainable and eco-friendly alternatives for the mitigation of hydrocarbon contaminants. The term "paraffin" can be used to describe a class of saturated hydrocarbon molecules known as the alkanes in chemistry and petro chemistry. Their straight-chain or branched-chain structure distinguishes these hydrocarbons, which have the general formula C_nH_{2n+2}. The separation of paraffins during the refining process results in a variety of petroleum products, which are frequently found in crude oil (Al-Yaari 2011). It can solidify at lower temperatures and form deposits in production equipment, leading to reduced oil flow rates and increased maintenance costs. Paraffin deposition is a widespread issue in the petroleum industry and requires continuous monitoring and management to prevent severe operational disruptions. Chemical treatments might be necessary to reduce paraffin-related issues. These chemicals could have an impact on the environment and raise production costs overall. Waxes are long-chained hydrocarbons that often contain straight or branched alkanes. At room temperature, they are solid, but when the temperature drops, they can crystallize from hydrocarbon fluids (Elkatory et al. 2022). Reservoirs of crude oil and natural gas inherently include waxes. Depending on the type of gas or oil field, their concentration can change. As a result of temperature changes during the transportation of crude oil or natural gas through pipelines, wax can accumulate on the inside surfaces of the pipe. This process, called wax deposition, may cause flow constraints and operational problems (Olajire 2021). High molecular weight, aromatic hydrocarbons known as asphaltenes are normally solid at room temperature but soluble in crude oil. Their precise chemical makeup can change depending on the source of the crude oil, but they all contain carbon, hydrogen, and oxygen, nitrogen, and sulfur atoms (Zheng et al. 2020a). They are insoluble in most solvents but can be dispersed in oil as colloidal particles. Asphaltenes have been known to cause issues such as emulsion formation, equipment fouling, and reservoir plugging. Their presence can significantly affect the quality and value of crude oil, making their removal or degradation essential for efficient oil production and processing. In recent years, the use of recyclable environmental-friendly fungi has emerged as a promising approach for the degradation of paraffin, waxes, and asphaltenes in the oil and gas industry (Falode and Olusegun 2006). In order to convert complex hydrocarbons like waxes, paraffin, and asphaltenes into simpler molecules, fungi have the ability to release enzymes. The accumulation of these compounds in pipes and equipment can be decreased through a process called as biodegradation. In order to create biological consortia that efficiently break down hydrocarbons, fungi can collaborate with other microorganisms like bacteria. This strategy could improve overall degrading effectiveness and environmental adaptability (Mohammadi et al. 2020).

8.1.1 Environmental Concerns and Challenges

The presence of paraffin, waxes, and asphaltenes in the oil and gas industry raises several environmental concerns and challenges. These compounds can have detrimental effects on ecosystems, human health, and the overall sustainability of the industry. Understanding and addressing these concerns is crucial for minimizing environmental impacts and ensuring responsible resource extraction.

1. *Oil spills*: Ecosystems can be severely harmed by unintentional spills of crude oil or petroleum products, which may contain waxes and asphaltenes. Oil spills have the potential to taint bodies of water, destroy aquatic life, and harm shorelines. (Gorbatiuk and Pasichna 2020).
2. *Bioaccumulation*: A bioaccumulation of oil and its components in species at higher trophic levels can result from oil entering the food chain. For humans and apex predators in impacted environments, this may pose health problems. (Almeda et al. 2013).
3. *Habitat Destruction*: Wetlands and coastal areas are just two examples of habitats that can be destroyed by an accumulation of oil, wax, and asphaltenes. Local wildlife populations may be dispersed and suffer a decline as a result of this. (Haider et al. 2021).
4. *Air and water Pollution*: The burning of hydrocarbons, which releases pollutants into the atmosphere, occurs often in the oil and gas business. This may result in poor air quality, smog, and the emission of greenhouse gases that fuel climate change and accumulation of these substances in water bodies lead to the water pollution (Ragothaman and Anderson 2017).
5. *Health and Safety Risks*: Exposure to hydrocarbons, including paraffin, waxes, and asphaltenes, can pose health risks to workers in the oil and gas industry. Inhalation, dermal contact, and ingestion of these substances may lead to respiratory issues, skin disorders, and other health complications (Liu et al. 2022).

Addressing these environmental concerns and challenges (Table 8.1) requires the development and implementation of sustainable and eco-friendly practices in the oil and gas industry. The use of recyclable environmental-friendly fungi for the degradation of paraffin, waxes, and asphaltenes offers a promising approach to mitigate these environmental impacts.

Table 8.1 Key environmental impacts and challenges associated with the oil and gas industries

Environmental impact	Description
Air pollution	Emissions of greenhouse gases and air pollutants from combustion processes, flaring, and venting
Water contamination	Discharge of produced water, drilling fluids, and chemical additives into water bodies
Soil contamination	Spills, leaks, and improper disposal of oil-based fluids and waste materials
Biodiversity loss	Habitat destruction, fragmentation, and disturbance from infrastructure development and oil spills
Climate change	Contribution to global warming through the release of greenhouse gases, particularly carbon dioxide and methane
Ecosystem disruption	Alteration of natural ecosystems, including wetlands and marine environments
Noise pollution	Noise generated from drilling, extraction, and transportation activities
Waste generation	Generation of solid waste, drilling cuttings, and other waste materials
Community health impacts	Potential risks to human health from exposure to pollutants and chemicals

8.1.2 Importance of Sustainable and Eco-Friendly Solutions

It is impossible to overestimate the significance of environmentally sound and sustainable solutions for the breakdown of paraffin, waxes, and asphaltenes in the oil and gas sector. If not handled properly, these chemical substances can lead to serious problems and environmental impact. Here, we delve into the importance of sustainable and eco-friendly solutions in the oil and gas industry as follows:

1. *Environmental Impact*: Pipelines, wellbore plugs, and processing machinery can deposit paraffin, wax, and asphaltenes, which can cause clogs and lower flow rates. Chemical solvents and mechanical cleaning, which are common techniques of removal, have the potential to release hazardous compounds into the environment. Sustainable solutions try to reduce or even reverse these harmful environmental effects. Sustainable and eco-friendly practices in the oil and gas industry are vital for minimizing environmental degradation. By reducing the release of pollutants, such as greenhouse gases and toxic substances, these solutions help mitigate climate change, preserve ecosystems, protect biodiversity, and safeguard water resources (Tarmizi and Brahmana 2023).
2. *Resource Recovery*: Businesses that invest in environmentally friendly and sustainable solutions are more likely to spur innovation and achieve a competitive edge in their sector. This may result in enhanced market positioning and company expansion. (Ekins and Zenghelis 2021).

3. *Reputation and Stakeholder Engagement*: Sustainable practices enhance the reputation of oil and gas companies, promoting positive stakeholder engagement. Stakeholders, including investors, customers, communities, and NGOs, increasingly value environmental responsibility. By adopting eco-friendly solutions, companies demonstrate their commitment to sustainability, which can lead to increased investor confidence, customer loyalty, and community support (Fallah Shayan et al. 2022).
4. *Cost Reduction and Efficiency*: Sustainable practices often lead to cost savings and operational efficiencies. Energy conservation, waste reduction, and the use of renewable energy sources not only reduce environmental impacts but also lower energy costs and improve resource utilization. Sustainable solutions can drive innovation, optimize processes, and enhance overall operational performance (Wan Ahmad et al. 2016).
5. *Long-Term Resilience and Adaptability*: The transition towards sustainability is not only essential for mitigating environmental risks but also for ensuring the long-term resilience and adaptability of the oil and gas industry. By embracing renewable energy sources, carbon capture and storage technologies, and circular economy principles, companies can position themselves for a future with evolving energy demands and stricter environmental constraints (Fraccascia et al. 2021).
6. *Social Responsibility and License to Operate*: The oil and gas industry operates within societies and communities that expect responsible behavior. Sustainable and eco-friendly solutions demonstrate a commitment to social responsibility, improving the industry's license to operate. This fosters positive relationships with local communities, Indigenous groups, and other stakeholders, thereby reducing social conflicts and enhancing overall industry acceptance (Frynas 2009).

The use of biodegradable chemicals, enzymatic treatments, nanotechnology, and environmentally friendly thermal techniques are a few examples of sustainable and environmentally friendly solutions for the breakdown of paraffin, wax, and asphaltene in the oil and gas sector. The industry may lessen its environmental impact and secure a more sustainable future by embracing these alternatives.

8.2 Fungi in Bioremediation

8.2.1 Role of Fungi in Environmental Cleanup

Fungi play a significant role in environmental cleanup and remediation due to their unique metabolic capabilities and ability to degrade a wide range of pollutants. They have attracted considerable attention as potential bioremediation agents for various contaminants, including hydrocarbons, heavy metals, pesticides, and pharmaceuticals. Understanding the role of fungi in environmental cleanup is crucial for developing effective and sustainable solutions to mitigate pollution. Here, we delve into the role of fungi in environmental cleanup as follows:

1. *Biodegradation of Organic contaminants*: Organic materials can be effectively broken down by fungi. Enzymatic processes enable them to convert complex organic contaminants like hydrocarbons, insecticides, and petroleum chemicals into less harmful, simpler compounds. For instance, some fungi, such as white rot fungus, are capable of degrading lignin, a complex organic substance present in wood, paper, and plant materials (Bala et al. 2022).
2. *Biotransformation*: Some poisonous compounds can undergo transformation by fungi into less damaging versions. Toxic chlorinated organic molecules, for instance, can be changed by some fungus into non-toxic chloride ions and carbon dioxide (Komorowicz et al. 2023).
3. *Mycoremediation of Pesticides and Pharmaceuticals*: Fungi exhibit diverse enzymatic activities that can break down a wide range of pesticides and pharmaceutical compounds. They produce enzymes, such as laccases, peroxidases, and cytochrome P450s, which can degrade these pollutants through enzymatic transformations (Spina et al. 2018). Fungal-based approaches have shown promise in the removal of pesticides and pharmaceutical residues from soils and water environments.
4. *Bioaccumulation and Biomagnification control*: Contaminants can be stopped by fungi before they reach the food chain. Fungi can lessen the risk of biomagnification, in which harmful compounds become more concentrated as they ascend up the food chain, by collecting and immobilizing contaminants in their mycelium (Miller et al. 2020).
5. *Phytoremediation Enhancement*: *Mycorrhizae*, which are advantageous symbiotic interactions between fungi and plants, increase the plants' capacity to absorb nutrients and tolerate pollutants (Rubin and Görres 2020). By using plants to remove toxins, phytoremediation can become more effective overall as a result.

It is significant to remember that the efficiency of fungus in environmental cleanup depends on a variety of elements, including the nature of the contaminants present, the individual fungi species used, the environmental circumstances, and the overall remediation plan. Fungi do, however, have a lot of promise for environmentally acceptable, long-lasting solutions to the problems of environmental contamination and repair.

8.2.2 Advantages of Fungi Over Other Microorganisms

Fungi possess several advantages over other microorganisms in various environmental applications, including bioremediation and industrial processes. These advantages stem from their unique physiological and metabolic characteristics, which make them versatile and effective in addressing environmental challenges. Here, we discuss the advantages of fungi over other microorganisms:

1. *Broad Substrate Range*: Fungi exhibit a remarkable ability to degrade diverse organic compounds, including complex polymers, hydrocarbons, and recalcitrant pollutants. They produce a wide array of extracellular enzymes capable of breaking down complex substrates, allowing them to utilize a broad range of carbon sources (Bianco and Perrotta 2015). This versatility makes fungi well-suited for the degradation of complex contaminants, such as crude oil, pesticides, and industrial wastes.
2. *Tolerance to Harsh Environmental Conditions*: Fungi are known for their resilience in harsh environmental conditions, including extremes of pH, temperature, and salinity. They can thrive in contaminated environments that may be inhibitory to other microorganisms (Gadd 2001). Fungal tolerance to adverse conditions enables them to survive and function effectively in polluted ecosystems, making them valuable for bioremediation applications.
3. *Production of Extracellular Enzymes*: Fungi are prolific producers of extracellular enzymes, including ligninolytic enzymes (e.g., lignin peroxidase, manganese peroxidase) and hydrolytic enzymes (e.g., cellulases, lipases). These enzymes have high catalytic activity and play a crucial role in the breakdown of complex organic compounds (Kaskatepe and Yildiz 2016). The ability of fungi to produce and secrete a diverse range of enzymes contributes to their effectiveness in degrading pollutants and recalcitrant substrates.
4. *Biofilm Formation and Attachment*: Fungi have the capability to form robust biofilms on various surfaces, allowing them to colonize and degrade pollutants in complex matrices, such as soils and sediments. Biofilm formation provides protection against adverse environmental conditions, enhances pollutant uptake, and promotes intercellular communication, leading to enhanced bioremediation efficiency (Saleem and Moe 2014). The ability of fungi to form biofilms contributes to their resilience and persistence in contaminated environments.
5. *Mycorrhizal Symbiosis and Plant Interactions*: Many fungi establish mutualistic relationships with plants through mycorrhizal associations. These symbiotic interactions enhance nutrient uptake, promote plant growth, and confer tolerance to environmental stresses. In the context of bioremediation, mycorrhizal fungi can facilitate the uptake and accumulation of contaminants in plants, improving phytoremediation efficacy (Dowarah et al. 2021). The ability of fungi to establish beneficial associations with plants provides additional benefits in environmental applications.

8.2.3 Types of Fungi Used in Bioremediation

Fungi are diverse microorganisms that exhibit immense potential for bioremediation due to their ability to degrade a wide range of contaminants. Different types of fungi have been employed in bioremediation processes, targeting specific pollutants and environmental conditions. Here, we discuss the types of fungi commonly used in bioremediation:

1. *White-Rot Fungi*: Lignin can be broken down into simpler molecules by ligninases, which are produced by white rot fungus and include lignin peroxidases, manganese peroxidases, and laccases. Lignin makes up a sizeable portion of woody plant matter and is extremely resistant to degradation. White rot fungus open up plant material's cellulose and hemicellulose for microbial destruction by dissolving lignin, which can help organic pollutants decompose. Additionally, lignin-degrading enzymes produced by white rot fungi can be used to degrade a variety of organic contaminants with lignin-like chemical compositions. This includes the dangerous and long-lasting chemical molecules known as PAHs, which are frequently discovered in tainted soil and water. Through the process of biodegradation, the enzymes created by white rot fungi can change these contaminants into less hazardous or non-toxic compounds (Suryadi et al. 2022).
2. *Brown-Rot Fungi*: The degradation of wood and lignocellulosic materials is one area in which brown rot fungi, like white rot fungi, have special properties that make them useful in bioremediation efforts. Brown rot fungi have developed unique enzymatic systems that make them well-suited for the breakdown of cellulose, a crucial component of plant cell walls, in contrast to white rot fungi, which largely break down lignin. Despite not having as much lignin-degrading power as white rot fungus, brown rot fungi can nevertheless help some organic contaminants found in wood or lignocellulosic materials biodegrade. This may involve the degradation of particular pesticide classes or chemical substances (Bugg et al. 2011).
3. *Mycorrhizal Fungi*: In a mutualistic connection, mycorrhizal fungus and plants trade nutrients (usually phosphorus and nitrogen) for carbohydrates that the plant produces during photosynthesis. A successful phytoremediation may depend on better plant development, which is frequently the result of this relationship. Additionally, certain pollutants might be better absorbed and accumulated by plants thanks to mycorrhizal fungus. This can be helpful in phytoremediation techniques, which concentrate contaminants in plant tissues for later harvest or disposal in order to remove them from the soil (Hakeem et al. 2014).
4. *Filamentous Fungi*: Filamentous fungi, including species from the genera *Aspergillus*, *Penicillium*, and *Fusarium*, have been widely utilized in various bioremediation applications. These fungi possess versatile metabolic capabilities and produce a range of extracellular enzymes that can degrade diverse organic pollutants. They have been employed in the bioremediation of contaminants such as hydrocarbons, pesticides, dyes, and pharmaceuticals (Ngo and Tischler 2022). Filamentous fungi offer advantages such as rapid growth, high enzyme production, and adaptability to different environmental conditions.
5. *Yeasts*: Certain yeast species, such as *Candida*, *Rhodotorula*, and *Saccharomyces*, have demonstrated potential in bioremediation processes. Yeasts are known for their robustness, adaptability, and ability to tolerate a wide range of environmental conditions. They can utilize various carbon sources, including pollutants like

hydrocarbons and aromatic compounds, and produce enzymes that facilitate the degradation of these contaminants (Gavrilescu 2010). Yeasts have been explored for their applications in the bioremediation of oil spills, wastewater treatment, and industrial waste remediation.

8.3 Understanding Paraffin, Waxes, and Asphaltenes

8.3.1 Properties and Composition of Paraffin

Paraffin is a type of hydrocarbon that consists primarily of saturated alkanes. It is a waxy, solid substance that is commonly found in petroleum and natural gas deposits. In addition to being used as a fuel and lubricant, paraffin is a form of hydrocarbon compound that is frequently used to make candles and other products that are made of wax. A wide range of substances belonging to the alkane family can be referred to as paraffin. Depending on the precise type and grade of paraffin, its characteristics and composition can change. The properties and composition of paraffin can vary depending on its source and specific chemical composition. The general properties and composition of paraffin are given below:

1. *Molecular Structure*: Paraffin is composed of long-chain hydrocarbons, specifically alkanes, which are characterized by a carbon backbone with single bonds and saturated with hydrogen atoms (Speight 2014). The general formula for paraffin is C_nH_{2n+2}, where n represents the number of carbon atoms in the molecule. The carbon atoms in paraffin chains are typically arranged in a linear or branched configuration.
2. *Physical State*: Paraffin exists as a solid at room temperature, with a waxy texture and appearance. The melting point of paraffin can vary depending on its specific composition, but it generally ranges from around 40 °C to 70 °C (Freund et al. 1982). The solid-state of paraffin enables its use in various applications, including candles, coatings, and as a phase-change material It typically has no smell and no color. Water does not dissolve paraffin, but many organic solvents do, including acetone, ether, and benzene.
3. *Thermal Stability*: Paraffin exhibits a high degree of thermal stability, making it resistant to decomposition and combustion at elevated temperatures. This property is attributed to the strong carbon–carbon and carbon-hydrogen bonds present in the paraffin structure. It also contributes to the high energy content of paraffin, making it a valuable source of fuel. Its high latent heat of fusion is one of paraffin wax's most noteworthy thermal characteristics. This is the quantity of heat energy necessary to transform a substance from a solid to a liquid at its melting point without a rise in temperature. Due to the relatively high latent heat of fusion of paraffin wax, it may store and release large amounts of energy during phase transitions.

4. *Chemical Inertness*: Paraffin is chemically inert and non-reactive with many substances, including water, acids, and bases (Freund et al. 1982). This inertness makes paraffin relatively stable and unreactive in various environments. However, certain conditions, such as exposure to strong oxidizing agents or UV radiation, can lead to chemical reactions or degradation of paraffin.
5. *Composition and Impurities*: The composition of paraffin can vary depending on its source and refining processes. In addition to the main alkane components, paraffin can contain impurities such as aromatic hydrocarbons, sulfur compounds, and trace metals. The impurities in paraffin can affect its properties and suitability for different applications (Britannica of Encyclopaedia 2015).

Paraffin wax is a versatile substance for a variety of applications, especially those involving heat storage, insulation, and controlled heat release, thanks to its thermal qualities overall. In many fields, such as thermal management systems, candle making, and others, its capacity to store and release energy during phase shifts is an important property.

8.3.2 Characteristics of Waxes in Oil and Gas Industries

Waxes are complex mixtures of long-chain hydrocarbons that are commonly encountered in the oil and gas industry. They can be found in crude oil, natural gas, and petroleum products, and their presence poses challenges in various operations. In the oil and gas sector, waxes are hydrocarbon molecules that can harden and precipitate at lower temperatures, posing problems for operations and production. In the context of the industry, it is crucial to comprehend certain features of these waxes. Here, we discuss the characteristics of waxes in the oil and gas industries:

1. *Solid State at Ambient Temperatures*: One of the primary characteristics of waxes is their solid state at ambient temperatures. Waxes have high melting points and exist as solid materials under normal operating conditions. The melting points of waxes can vary depending on their comp8sition, ranging from approximately 40 °C to 90 °C (Machado et al. 2001). The solid nature of waxes can lead to the formation of deposits and blockages in pipelines, equipment, and storage facilities, causing operational issues.
2. *High Molecular Weight and Long-Chain Hydrocarbons*: Waxes are composed of high molecular weight hydrocarbons, typically consisting of long-chain alkanes, branched alkanes, and cycloalkanes. The length of the carbon chains in waxes can range from around 20 to 60 carbon atoms (Leontaritis 1996). The presence of these long-chain hydrocarbons contributes to the solid and viscous nature of waxes.
3. *Crystalline Structure*: Waxes exhibit a crystalline structure, which further enhances their solid-state behavior. The molecules in waxes arrange themselves in an ordered pattern, forming crystalline structures. The specific crystalline structures can vary depending on the composition and cooling rate during wax

formation. The crystal lattice of waxes contributes to their physical properties, such as brittleness, hardness, and low solubility in common solvents (Leontaritis 1996).

4. *Solubility in Non –polar solvents*: Alkanes and mineral oils are examples of non-polar solvents where paraffinic waxes can be dissolved, despite the fact that they are often insoluble in water. In the oil and gas business, this feature may be used in the creation of wax inhibitors and removal techniques (Wang et al. 2022).
5. *Role in Wax Deposition*: Wax deposits in pipelines and other oil and gas sector equipment are largely the result of waxes' paraffinic composition. Paraffinic waxes have a tendency to crystallize and create solidified formations that can build up and block flow as the temperature of crude oil or hydrocarbons falls during processing or transportation (Chala et al. 2018).

8.3.3 Introduction to Asphaltenes and Their Impact on Oil and Gas Operations

Asphaltenes are complex organic compounds that occur naturally in crude oil and can have a significant impact on oil and gas operations. They are a high molecular weight fraction of petroleum and are composed of a mixture of aromatic and aliphatic hydrocarbons, along with heteroatoms such as nitrogen, oxygen, and sulfur. Asphaltenes are intricate, high-molecular-weight hydrocarbon molecules that are present in crude oil and natural bitumen. They make up a sizeable portion of the non-volatile portion of these materials and are renowned for their complex and polymeric makeup.

1. *Definition and Occurrence*: Asphaltenes are defined as the fraction of petroleum that is insoluble in n-pentane, but soluble in aromatic solvents such as toluene or xylene. They are typically present in crude oil at concentrations ranging from 0.5% to 20% or higher, depending on the oil reservoir (Vargas et al. 2014). Asphaltenes can also be found in natural gas, where they may precipitate due to changes in pressure, temperature, or composition.
2. *Physical and Chemical Properties*: Asphaltenes exhibit a range of physical and chemical properties. They are dark, solid materials with a high molecular weight and complex molecular structure. The molecular weight of asphaltenes can vary from several thousand to tens of thousands (Zheng et al. 2020b). Asphaltenes possess high aromaticity, with a significant number of fused aromatic rings in their structure, which contributes to their complex and polydisperse nature. Asphaltenes may increase oil viscosity and decrease fluidity in pipelines and production systems, resulting in flow limits and challenges with processing and transportation.
3. *Challenges in Oil and Gas Operations*: Asphaltenes pose several challenges in oil and gas operations. They can precipitate and form solids or deposits under certain conditions, leading to flow restrictions, plugging of reservoirs, and fouling of production equipment (Speight and (Ed.). 1999). Asphaltenes can also cause emulsion formation, stability issues, and corrosion in pipelines and processing

facilities. These challenges can have a negative impact on production rates, increase operational costs, and require additional treatments for oil recovery and processing.

4. *Impact on Refining*: Asphaltenes can have a significant impact on petroleum refining processes. During crude oil distillation, asphaltenes tend to remain in the residue, which affects the quality and yield of refined products. They can contribute to the formation of coke and fouling in refining units, reduce catalyst activity, and affect the performance of downstream processes such as hydrotreating and catalytic cracking (Idris and Okoro 2013). Proper management of asphaltenes is essential for efficient refining operations.
5. *Mitigation and Characterization*: Understanding and mitigating the challenges associated with asphaltenes is an active area of research. Various techniques are employed to characterize asphaltenes, including spectroscopic methods, molecular modeling, and separation techniques (Seitmaganbetov et al. 2021). Developing effective mitigation strategies such as asphaltene inhibitors, dispersants, and solvents is crucial to prevent or reduce the adverse effects of asphaltenes on oil and gas operations.

Therefore, Asphaltenes are intricate, high-molecular-weight substances that are present in bitumen and crude oil. They are recognized for their insolubility, tendency to aggregate, and influence on fluid characteristics. To maintain the smooth flow of hydrocarbons and avoid operational difficulties, managing asphaltenes is a crucial component of oil production and processing.

8.4 Mechanisms of Fungal Degradation

8.4.1 Enzymatic Breakdown of Paraffin and Waxes by Fungi

Fungi play a crucial role in the degradation of paraffin and waxes through enzymatic processes. They produce a variety of enzymes capable of breaking down the complex hydrocarbon structures present in paraffin and waxes. The enzymatic breakdown of paraffin and waxes by fungi involves a series of steps that are facilitated by specific enzymes produced by the fungi which are shown in (Fig. 8.1) and discussed below:

1. *Surface Attachment and Emulsification*: Fungi first attach to the surface of the paraffin or wax material using their mycelia, which are branching filaments. The mycelia produce extracellular enzymes, including lipases and esterases, which act on the hydrocarbon chains of the paraffin and waxes. These enzymes help in emulsifying the hydrophobic material, breaking it down into smaller droplets or micelles (Lv et al. 2023).
2. *Enzymatic Hydrolysis*: The lipases and esterases produced by fungi play a key role in the enzymatic breakdown of paraffin and waxes. Lipases hydrolyze the ester bonds present in the hydrocarbon chains of paraffin and waxes, releasing

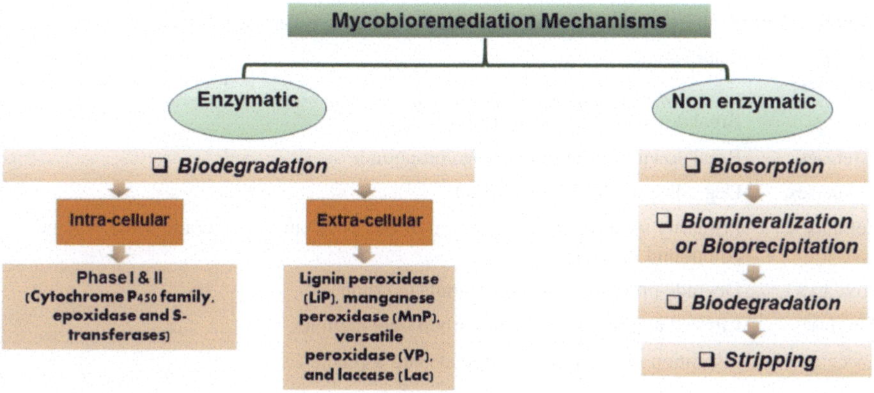

Fig. 8.1 Mechanisms involve in myco-bioremediation

fatty acids and alcohols. Esterases, on the other hand, hydrolyze the ester bonds in the wax esters, leading to the formation of fatty acids and alcohols (Cong et al. 2019). These enzymatic reactions result in the breakdown of the complex hydrocarbon structures into smaller, more metabolizable compounds.

3. *Further Metabolism*: Once the hydrocarbons are converted into fatty acids and alcohols, fungi utilize these smaller compounds as carbon and energy sources for their growth and metabolism. The fatty acids and alcohols are taken up by the fungal cells and metabolized through various pathways, such as β-oxidation and tricarboxylic acid cycle, to generate energy and biomass (Cong et al. 2019).
4. *Synergistic Action*: The enzymatic breakdown of paraffin and waxes by fungi often involves a synergistic action of multiple enzymes. Fungi produce a variety of enzymes, including lipases, esterases, and oxidative enzymes, which work together to efficiently break down the complex hydrocarbon structures. The combination of these enzymes allows fungi to degrade paraffin and waxes and utilize them as a carbon and energy source (Gadd and (Ed.). 2006).

8.4.2 Fungal Metabolites and Their Role in Asphaltenes Degradation

Fungi produce a diverse array of secondary metabolites, including enzymes, organic acids, and bioactive compounds, that play a crucial role in the degradation of asphaltenes. These fungal metabolites contribute to the breakdown and transformation of complex hydrocarbon structures, enabling the utilization of asphaltenes as a carbon and energy source. In this section, we discuss the role of fungal metabolites in asphaltenes degradation (Table 8.2).

1. *Enzymes*: Fungi produce a range of extracellular enzymes that are involved in the degradation of asphaltenes. These enzymes include oxidases, peroxidases,

Table 8.2 List of fungal enzymes used for the degradation of waxes, paraffin and asphaltenes

Fungal Enzymes	Application
Lipases	Breakdown of waxes and triglycerides
Esterases	Breakdown of ester containing compounds
Xylanases	Degrade complex Hydrocarbons
Cellulases	Primarily used for the degradation of cellulose but also degrade some organic compound in crude oil
Proteases	Degrade proteinaceous compounds in crude oil
Amylases	Limited application in hydrocarbon degradation
Laccases	Oxidation of phenolic compounds and asphaltenes

and laccases, which participate in oxidative reactions and modify the structure of asphaltenes. Oxidative enzymes can catalyze the oxidation of asphaltenes, breaking down aromatic rings and generating smaller, more metabolizable fragments. Wax and triglycerides can be hydrolyzed by the enzymes known as lipases (Imam et al. 2019). The ability of several fungal lipases to degrade waxy components in crude oil has been investigated. The hydrolysis of ester bonds is facilitated by enzymes called esterases. They may be involved in the breakdown of chemicals found in crude oil that contain esters, such as wax esters. Several chemical molecules, including some parts of asphaltenes, can be degraded by laccases, oxidative enzymes. They take part in the oxidation of phenolic substances. The action of these enzymes is crucial for initiating the degradation process of asphaltenes by fungi.

2. *Organic Acids*: Fungi also secrete organic acids, such as citric acid, oxalic acid, and acetic acid, which contribute to the solubilization and breakdown of asphaltenes (Gao et al. 2017). Organic acids can chelate metal ions present in asphaltenes, disrupting their structure and enhancing their solubility. Moreover, organic acids can act as electron shuttles, facilitating redox reactions and promoting the activity of oxidative enzymes involved in asphaltenes degradation. It's crucial to keep in mind that the precise kinds and quantities of organic acids produced can vary depending on the fungus species and the environmental circumstances in which they are flourishing. To convert complicated hydrocarbons like waxes and asphaltenes into simpler, easier-to-digest molecules, fungi frequently use a combination of enzymes and organic acids. The carbon and nitrogen cycling of the ecosystem includes this deterioration as a key component.
3. *Bioactive Compounds*: Fungal metabolites, including bioactive compounds, can have an indirect impact on asphaltenes degradation. These bioactive compounds may include antibiotics, siderophores, and secondary metabolites with growth-promoting or inhibitory effects on other microorganisms present in the environment (Al-Rashdi et al. 2023). By affecting the microbial community dynamics, fungal bioactive compounds can influence the overall efficiency of asphaltenes degradation.

4. *Synergistic Interactions*: Fungal metabolites often work synergistically to enhance the degradation of asphaltenes. The combination of enzymes, organic acids, and bioactive compounds produced by fungi can create an environment conducive to asphaltenes breakdown. Synergistic interactions among these metabolites can promote the solubilization, modification, and utilization of asphaltenes as a carbon source (Malik 2008).
5. *Biotechnological Applications*: Understanding the role of fungal metabolites in asphaltenes degradation has significant biotechnological implications. Harnessing the enzymatic and chemical activities of fungal metabolites can aid in the development of efficient bioremediation strategies for petroleum-contaminated environments. Additionally, exploring fungal metabolites opens avenues for the production of value-added products from asphaltenes, such as biofuels or bio-based chemicals (Malik 2008).

8.4.3 Key Factors Affecting Fungal Degradation Efficiency

The efficiency of fungal degradation in environmental cleanup processes depends on various factors that influence the growth and activity of fungi. Understanding these key factors is essential for optimizing fungal degradation efficiency and developing effective bioremediation strategies. Some of the key factors that affect fungal degradation efficiency are discussed below:

1. *Environmental Conditions*: Fungal degradation efficiency is greatly influenced by environmental factors such as temperature, pH, oxygen availability, and nutrient availability. Temperature has a significant impact on fungal activity. The majority of fungi have a temperature range where growth and enzyme activity are ideal. Extreme temperatures, whether they are too high or low, can prevent the decomposition of fungi and for optimum development and enzyme activity, fungi have certain pH requirements. Efficiency of degradation may be impacted by pH variations from the desired range. For fungi to grow and create enzymes, nutrients including carbon, nitrogen, and phosphorus are essential. The substrate's ability to contain these nutrients can have a big impact on how well things degrade. The majority of fungi are aerobic, which means they need oxygen for metabolism. Changes in oxygen concentrations can affect the kinds of degradation pathways fungi employ and how effective they are in general (Maheshwari et al. 2000).
2. *Substrate Compatibility*: The type and composition of the substrate or contaminant being degraded can significantly impact fungal degradation efficiency. Different fungi have specific substrate preferences and abilities to degrade various contaminants. For example, certain fungi may have a higher affinity for petroleum hydrocarbons, while others are more efficient at degrading specific compounds like polycyclic aromatic hydrocarbons (PAHs) or chlorinated pollutants. Matching the right fungal species to the specific substrate or contaminant can enhance degradation efficiency (Jeon et al. 2016). Additionally, the physical form of the substrate, such as its solubility, accessibility, and surface area, can

also influence fungal degradation efficiency, hence, the effectiveness of fungal degradation can be significantly impacted by the type of organic substance being broken down. Certain fungi can degrade some substrates more readily than others, while others may be more resistant.

3. *Competition*: For the same organic material as a carbon and energy source, fungi, bacteria, and other microbes fight each other. It can slow down fungal growth and deterioration when several species compete for the same scarce nutrients in a substrate. For example, bacteria have a higher rate of compound consumption than fungi, which leaves fungi with fewer resources to operate with. In addition to needing carbon to thrive and create enzymes, fungi also need nitrogen, phosphorus, and other nutrients. For certain nutrients, bacteria can outcompete fungi, which can reduce the effectiveness of fungal breakdown. Some bacteria can create substances that prevent the development of fungal spores, which hinders the ability of the fungi to reproduce and colonize new places. Understanding and controlling microbial competition is crucial for optimizing fungal degradation processes in applications including bioremediation, composting, and waste management. This may entail choosing particular fungal strains or altering the environment to favor the appropriate microbial population and increase degrading effectiveness (Gupta et al. 2016).
4. *Co-Contaminant Interactions*: The presence of co-contaminants in the environment can impact fungal degradation efficiency. Some co-contaminants may inhibit fungal growth or interfere with degradation processes, while others may have a synergistic effect. Toxins or other contaminants in the substrate can prevent microbial growth and enzymatic activity, which lowers the effectiveness of the degradation process. For example, heavy metals and certain organic compounds can be toxic to fungi and hinder their degradation activities. On the other hand, the presence of certain co-contaminants, such as phenolic compounds or organic acids, can enhance fungal degradation efficiency by stimulating enzyme production or providing additional carbon sources (Zhang and Zhang 2022).
5. *Operational Strategies*: Operational strategies, such as the application method of fungal inoculants, duration of treatment, and the use of supportive technologies, can affect degradation efficiency. Proper distribution and contact between fungi and contaminants are crucial for effective degradation. Enhanced bioavailability through surfactant addition, bioaugmentation with adapted fungal strains, and the use of immobilized fungi can improve degradation efficiency and long-term stability (Atlas and Bragg 2009).

8.5 Fungal Strains for Paraffin, Waxes, and Asphaltenes Degradation

8.5.1 Selection Criteria for Fungal Strains

When considering fungal strains for the degradation of paraffin, waxes, and asphaltenes, certain selection criteria should be taken into account. These criteria ensure the effectiveness of the fungal degradation process and the ability of the selected strains to efficiently degrade the target hydrocarbon compounds. The key selection criteria for fungal strains are given below:

1. *Hydrocarbon-Degrading Capability*: Hydrocarbons, such as different kinds of petroleum hydrocarbons including crude oil, gasoline, and diesel fuel, have been proven to be capable of being degraded by fungi. Through a process known as biodegradation, they achieve this by using enzymes to disassemble hydrocarbon molecules into less complex chemicals that can serve as a carbon source for their growth. Different fungal strains have proven to be capable of decomposing hydrocarbons. *Aspergillus*, *Penicillium*, *Trichoderma*, *Fusarium*, *Candida*, and *Rhodococcus* are a few common genera with a reputation for having this ability. Different enzymes and methods for the breakdown of hydrocarbons may be present in each of these fungi. Numerous fungi may adapt to grow in hydrocarbon-contaminated environments. Changes in gene expression and the creation of particular enzymes that target hydrocarbons are frequent components of this adaptation (Daccò et al. 2020).
2. *Substrate Specificity*: Depending on the fungal species and its enzymatic capabilities, the substrate specificity of fungal strains for the breakdown of waxes, paraffin, and asphaltenes can change. Due to the differing chemical structures of these three hydrocarbons, which are found in crude oil and petroleum products, separate enzymes are needed for their breakdown.

   - **Waxes**

     Composition: The main constituents of waxes are long-chain saturated hydrocarbons, which are frequently found as esters or paraffins.
     Fungal Degradation: Esterases and lipases are frequently produced by fungi that specialize in the breakdown of wax. These enzymes can hydrolyze ester bonds in wax molecules, dissolving them into less complex fatty acids and alcohols that the fungus can use to thrive. *Aspergillus*, *Candida*, and *Trichosporon* are some fungal species that have been linked to wax degradation (Chandra et al. 2020).

   - **Paraffin**

     Composition: A form of hydrocarbon called paraffin is made up either straight-chain or branched alkanes.

Fungal Degradation: Enzymes like alkane monooxygenases or cytochrome p450 enzymes are frequently produced by fungi that may break down paraffin. The hydrocarbons can be oxidized by these enzymes, which makes them more receptive to further deterioration. Some hydrocarbon-degrading fungi, including specific strains of *Trichoderma*, are known for their capacity to break down paraffin (Das and Chandran 2011).

- **Asphaltene**

  Composition: Asphaltenes, which are composed of highly aromatic and heteroatom-rich molecules, are the heaviest and most complicated portion of crude oil.
  Fungal Degradation: Due to their complexity and resistance to fungus, asphaltenes are less frequently and more difficultly degraded. Some fungus, including some *Fusarium* species, have been found to contain enzymes that can degrade some asphaltenes' constituent parts. To completely degrade asphaltenes, however, a microbial consortium's combination of bacterial and fungal activity may be necessary (Al-Otibi et al. 2023).

3. *Tolerance to Environmental Conditions*: The selected fungal strains should exhibit tolerance to the environmental conditions of the specific site where the degradation process will take place. Factors such as temperature, pH, salinity, and presence of toxic compounds can influence fungal growth and degradation activity. Therefore, it is crucial to choose fungal strains that can thrive and perform effectively under the prevailing environmental conditions of the contaminated site. This ensures their survival and sustained activity during the degradation process (Deshmukh et al. 2016).
4. *Compatibility with Co-existing Microorganisms*: In many environmental settings, fungal degradation occurs alongside other microorganisms, including bacteria and other fungi. It is essential to consider the compatibility of the selected fungal strains with co-existing microorganisms. Some fungi may have synergistic interactions with certain bacteria, leading to enhanced degradation efficiency. Conversely, compatibility issues or antagonistic interactions between fungal and bacterial species may hinder the degradation process. Therefore, understanding the microbial community dynamics and considering the compatibility of fungal strains with co-existing microorganisms is crucial for successful degradation (Krüger et al. 2019).

8.5.2 Promising Fungal Species and Their Applications

Several fungal species have shown great promise in the degradation of paraffin, waxes, and asphaltenes in the oil and gas industries. These fungi possess specific characteristics and metabolic capabilities that make them suitable candidates for bioremediation and environmental cleanup applications. Some of the most promising fungal species and their applications are listed below (Table 8.3).

Table 8.3 Fungal species and their applications in hydrocarbon degradation

Fungal species	Applications	References
Trichoderma spp.	Biodegradation of paraffin, waxes, and asphaltenes	Ghosh and Khare (2016)
Aspergillus niger	Production of enzymes for hydrocarbon degradation	Mittal and Singh (2009)
Phanerochaete chrysosporium	Bioremediation of oil spills and hydrocarbon-contaminated soil	Xenia and Refugio (2016)
Penicillium spp.	Biodegradation of waxes in oil and gas pipelines	Nesaratnam (1994)
Candida tropicalis	Degradation of asphaltenes in heavy crude oil	He et al. (2014)
Bjerkandera adusta	Degradation of aromatic hydrocarbons and recalcitrant compounds	Kadri et al. (2017)
Cunninghamella elegans	Degradation of hydrocarbons and production of extracellular enzymes	Al-Hawash et al. (2018)
Mortierella alpine	Degradation of long-chain alkanes and polycyclic aromatic hydrocarbons	Chandra et al. (2013)
Fusarium spp.	Degradation of diesel fuel and petroleum-derived compounds	Fenibo et al. (2019)
Paecilomyces lilacinus	Degradation of petroleum hydrocarbons	Benguenab and Chibani (2021)
Pleurotus ostreatus	Biodegradation of hydrocarbons and wax esters	Tiwari et al. (2021)
Candida tropicalis	Degradation of asphaltenes in heavy crude oil	Harayama et al. (1999)
Phoma spp.	Biodegradation of petroleum hydrocarbons and wax	Ramdass and Rampersad (2021)
Cladosporium sphaerospermum	Degradation of alkanes and petroleum hydrocarbons	Naeem and Qazi (2020)
Metarhizium spp.	Biodegradation of hydrocarbons and asphaltenes	Aranda (2016)
Yarrowia lipolytica	Biodegradation of hydrocarbons and waxes	Harzevili (2014)
Myrothecium verrucaria	Degradation of crude oil and paraffin wax	Arunkumar et al. (2017)
Aureobasidium pullulans	Biodegradation of petroleum hydrocarbons and waxes	Kaplan (1998)
Curvularia spp.	Degradation of alkanes and polycyclic aromatic hydrocarbons	Khan et al. (2019)
Penicillium janthinellum	Biodegradation of paraffin and petroleum hydrocarbons	Khan et al. (2015)

8.5.3 Case Studies on Successful Fungal Remediation

Fungal remediation has shown great potential in the bioremediation of oil and gas-contaminated sites. Several case studies have demonstrated the successful application of fungi in degrading paraffin, waxes, and asphaltenes, leading to the restoration of contaminated environments. Here, we discuss two notable case studies that highlight the effectiveness of fungal remediation:

1. Case Study: Fungal Remediation of Crude Oil-Contaminated Soil

A case study conducted by Atagana et al. (2006) focused on the remediation of crude oil-contaminated soil using a fungal-based approach. In this study, three fungal species, *Trichoderma viride*, *Aspergillus niger*, and *Penicillium chrysogenum*, were evaluated for their ability to degrade hydrocarbons present in the contaminated soil. The fungal inoculum was applied to the soil, and various parameters, such as hydrocarbon degradation efficiency and microbial community dynamics, were monitored over time. The results of the study demonstrated the effectiveness of the fungal remediation approach. After a 90-day incubation period, the fungal-treated soil showed a significant reduction in total petroleum hydrocarbon (TPH) content compared to the untreated control soil. *Trichoderma viride* exhibited the highest degradation efficiency, with approximately 64% reduction in TPH content. The study also revealed changes in the microbial community structure, indicating the active involvement of fungi and other microorganisms in the degradation process.

2. Case Study: Fungal Remediation of Oil-Contaminated Water

A case study conducted by Cerniglia et al. (2001) focused on the remediation of oil-contaminated water using a fungal-based approach. In this study, a consortium of fungi, including *Trichoderma harzianum*, *Penicillium citrinum*, and *Aspergillus niger*, was applied to simulate oil-contaminated water samples collected from a spill site. The fungal consortium was evaluated for its ability to degrade the hydrocarbon components of the oil and reduce the overall toxicity of the contaminated water. The results of the study demonstrated the successful remediation of the oil-contaminated water by the fungal consortium. After a 14-day incubation period, the fungal-treated water samples showed a significant reduction in hydrocarbon content compared to the untreated samples. The fungal consortium effectively degraded various hydrocarbon fractions, including paraffin, waxes, and asphaltenes, leading to the detoxification of the water samples. This case study provides evidence of the effectiveness of fungal remediation in treating oil-contaminated water.

8.6 Optimization Techniques for Enhanced Degradation

8.6.1 Physicochemical Parameters for Optimal Fungal Growth

To maximize the efficiency of fungal degradation and enhance the breakdown of paraffin, waxes, and asphaltenes, it is important to optimize the physicochemical parameters that support fungal growth and activity. These parameters influence fungal metabolism, enzyme production, and overall degradation efficiency. Here, we discuss key physicochemical parameters that can be manipulated to create favorable conditions for fungal growth and enhanced hydrocarbon degradation.

1. *Temperature*: Temperature plays a crucial role in fungal growth and enzymatic activity. Different fungal species have specific temperature ranges in which they thrive. It is important to identify the optimal temperature for the selected fungal strain to maximize degradation efficiency. In general, mesophilic fungi prefer moderate temperatures between 25 °C and 35 °C, while thermophilic fungi perform best at higher temperatures between 45 °C and 55 °C (Balaji et al. 2014). Maintaining the appropriate temperature range ensures optimal fungal growth and metabolic activity, leading to efficient hydrocarbon degradation.
2. *pH*: Fungal growth and enzyme production are also influenced by the pH of the environment. Each fungal species has a specific pH range in which it exhibits optimal growth and enzyme secretion. It is important to adjust the pH of the contaminated site or the growth medium to match the preferred pH of the selected fungal strain. For example, some fungi thrive in acidic conditions, while others prefer neutral or slightly alkaline pH (Nagy et al. 2016). Providing the appropriate pH conditions promotes the growth of the selected fungal species and enhances their ability to degrade hydrocarbons.
3. *Nutrient Availability*: Fungal growth and metabolism are dependent on the availability of essential nutrients. These include carbon, nitrogen, phosphorus, and trace elements. Optimal nutrient availability is crucial for the synthesis of enzymes involved in hydrocarbon degradation. The carbon source, in particular, should match the hydrocarbon compounds present in the contaminated site. Providing the necessary nutrients in the right proportions promotes fungal growth and ensures efficient degradation of paraffin, waxes, and asphaltenes (Prince and Clark 2004).
4. *Oxygen Availability*: Fungi are aerobic organisms that require oxygen for their metabolic processes. Adequate oxygen supply is essential for optimal fungal growth and enzyme production. Proper aeration of the contaminated site or growth medium ensures sufficient oxygen availability to support fungal activities. In some cases, enhanced aeration techniques, such as air sparging or mechanical agitation, may be employed to improve oxygen transfer and promote fungal degradation efficiency (Boonchan et al. 2000).

5. *Co-substrates and Inducers*: Certain co-substrates and inducers can stimulate fungal enzyme production and enhance hydrocarbon degradation. Co-substrates, such as simple sugars or organic acids, can serve as additional carbon sources, promoting fungal growth and enzyme synthesis. Inducers, on the other hand, are specific compounds that trigger the production of target enzymes involved in hydrocarbon degradation. Manipulating the presence of co-substrates and inducers can enhance the enzymatic capabilities of fungi and improve their degradation efficiency (Yang and Zhang 2018).

8.6.2 Bioaugmentation and Biostimulation Strategies

Bioaugmentation and biostimulation are two strategies commonly employed in the bioremediation of oil and gas-contaminated environments. These approaches aim to enhance the degradation of paraffin, waxes, and asphaltenes by introducing specific microorganisms or providing favorable conditions for indigenous microbial communities. Here, we discuss these strategies in detail:

1. *Bioaugmentation*:

Bioaugmentation involves the addition of specific microorganisms, such as fungi, bacteria, or microbial consortia, to the contaminated site. The selected microorganisms are chosen for their ability to degrade hydrocarbons effectively. By introducing these specialized degraders, the microbial diversity and metabolic potential of the contaminated environment are enhanced, leading to accelerated hydrocarbon degradation. Enzymes that may degrade hydrocarbons, such as peroxidases and lipases, are produced by fungi. Enzyme production should be improved using techniques, including genetic engineering if necessary. To achieve this, the fungus strains may be altered to overexpress the necessary enzymes. Bioaugmentation has several advantages. It allows the introduction of highly efficient hydrocarbon-degrading microorganisms that may not be naturally present in the contaminated site. These introduced organisms can possess unique metabolic capabilities, including the ability to degrade complex hydrocarbon compounds. Bioaugmentation can also overcome the limitations of indigenous microbial populations, which may be hindered by factors such as nutrient limitations or toxic compounds. To ensure successful bioaugmentation, it is important to select the appropriate microorganisms based on their hydrocarbon degradation capabilities, compatibility with the contaminated site, and survival under site-specific conditions. Additionally, factors such as the timing and method of microbial inoculation should be carefully considered to maximize the effectiveness of the bioaugmentation process (Vásquez-Murrieta et al. 2016).

Wax, paraffin, and asphaltene bioremediation can be improved with the help of fungus, but this approach needs to be carefully planned, monitored, and optimized in order to be successful. It's also crucial to take into account the particular fungal strains and environmental circumstances that are particular to your cleanup site.

2. *Biostimulation*:

When compared to introducing non-native species, biostimulation takes advantage of the inherent capacities of native or introduced fungi to break down hydrocarbons. The appropriateness of the local fungal population and the improvement of site-specific conditions, however, are key factors in success. The goal of biostimulation is to promote the proliferation and metabolic activity of local fungi that already exist in contaminated environments. It may include site assessment, identify indigenous fungi their nutrient amendment and optimization conditions including pH adjustment, moisture content and aeration control.

Microbial enhanced oil recovery (MEOR) approaches use biostimulation to boost oil recovery from mature reservoirs. In order to change the oil's characteristics, lower viscosity, and enhance oil flow, microorganisms are added to the reservoir. Increased oil output may result from this. Biostimulation is employed in the oil and gas sector to remediate sour gas, which contains hazardous hydrogen sulfide. Biological desulfurization is a method for encouraging microbial communities to oxidize hydrogen sulfide to less harmful forms, including sulfate. Oilfield wastewater and generated water are treated using biostimulation. Water discharges can be made to comply with environmental laws by encouraging microbes to degrade organic pollutants and remove heavy metals (Haritash and Kaushik 2009). Pipelines can be filled with microorganisms that break down hydrocarbons to combine biostimulation and bioaugmentation. By controlling the buildup of paraffin and wax deposits, this method can lessen the need for chemical treatments or mechanical cleaning.

In order to address environmental and operational issues, use fewer chemicals, and encourage the responsible use of hydrocarbon resources, the oil and gas industry can benefit from biostimulation. But for it to be effective, each location must be carefully analyzed, and the best biostimulation methods must be chosen based on the unique problems at hand.

8.6.3 Integration of Fungal Remediation with Existing Oil and Gas Processes

The integration of fungal remediation with existing oil and gas processes offers several advantages, including cost-effectiveness, sustainability, and the potential for simultaneous waste management and resource recovery. By incorporating fungal remediation strategies into oil and gas operations, the degradation of paraffin, waxes, and asphaltenes can be enhanced, leading to improved environmental performance e and operational efficiency.

1 *Bioremediation in Produced Water Treatment*: Produced water is a significant waste stream generated during oil and gas production, and it often contains high levels of hydrocarbons and other contaminants. Fungal remediation can be integrated into the existing water treatment processes to enhance the removal of hydrocarbons from produced water. One approach is to utilize fungal bioreactors, where

the produced water is treated by passing it through a reactor containing fungal cultures capable of degrading hydrocarbons. The fungi metabolize the hydrocarbons present in the produced water, reducing their concentration to permissible levels. This integration not only facilitates the treatment and reuse of produced water but also helps in recovering valuable resources from the waste stream (Sokal et al. 2022).

2 *Bioaugmentation in Oilfield Soil and Sludge Treatment*: Oilfield soils and sludges are often contaminated with hydrocarbons and pose challenges for effective remediation. Integration of fungal remediation, specifically bioaugmentation, can enhance the degradation of paraffin, waxes, and asphaltenes in these matrices. By introducing hydrocarbon-degrading fungi into the contaminated soil or sludge, the indigenous microbial populations can be supplemented with specialized degraders, leading to accelerated hydrocarbon degradation. This integration can be carried out through the addition of fungal inoculants or by creating favorable conditions for the growth and activity of indigenous hydrocarbon-degrading fungi (Siles and García-Sánchez 2018).

The integration of fungal remediation with existing oil and gas processes presents opportunities for sustainable waste management and resource recovery. By leveraging the inherent capabilities of fungi in degrading hydrocarbons, the environmental impact of oil and gas operations can be minimized, and valuable resources can be recycled and reused.

8.6.4 Genetic Engineering for Improvement of Fungus Ability to Degrade Hydrocarbons

Hydrocarbons are organic substances that are present in petroleum-based goods like oil and gasoline, and genetic engineering can be used to improve the capacity of fungi to decompose them. Fungi are well known for their innate ability to degrade numerous organic compounds, including hydrocarbons, through enzymatic reactions. However, genetic engineering can increase their effectiveness (Dell' Anno et al. 2021).

Steps involved in enhancing fungus ability to degrade hydrocarbons are shown in Fig. 8.2.

Select a species of fungus that is naturally able to break down hydrocarbons. Oyster mushrooms (*Pleurotus ostreatus*) and white rot fungus (such as *Phanerochaete chrysosporium*) are examples of species that are well-known for their capacity to break down hydrocarbons. Analyze the genetic makeup of the chosen fungus to discover the genes and enzymes responsible for hydrocarbon breakdown. Understanding natural systems and routes will be aided by this. Identify the genes in the fungus' genome that are in charge of hydrocarbon breakdown. Molecular biology methods like PCR (Polymerase Chain Reaction) can be used for this. Improve the isolated genes' activity by making changes to them. This may entail numerous technologies, such as CRISPR-Cas9 gene editing or synthetic biology techniques. The

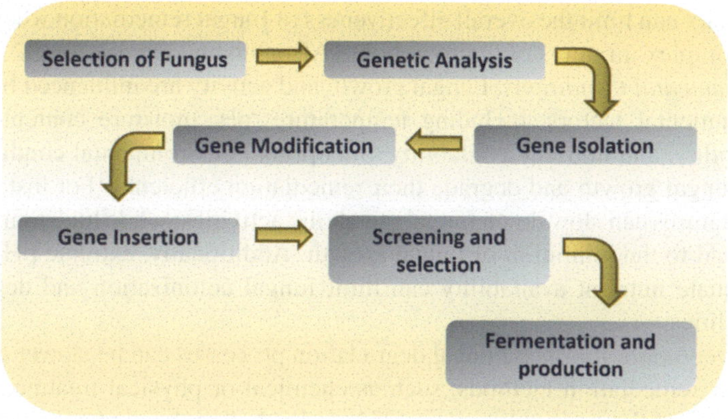

Fig. 8.2 General steps involved in enhancing fungus ability to degrade hydrocarbons

objective is to maximize the expression of genes that produce enzymes involved in the breakdown of hydrocarbons. Reintroduce the altered genes into the fungus' genome. Plasmids or other types of gene delivery systems are frequently used for this. Strong promoters should be in charge of the inserted genes to ensure maximum enzyme output. Grow the genetically altered fungus and test it for stronger hydrocarbon-degrading capacities. This can be accomplished by exposing the fungus to settings rich in hydrocarbons and tracking the rate of breakdown (Arun et al. , 2023).

When genetically altering fungi or any other organism for environmental applications, careful investigation and risk evaluations are essential. Before employing such bioremediation techniques on a wide scale, it is additionally necessary to acquire governmental and public acceptance. The objective is to utilize fungi's inherent qualities while reducing dangers and maximizing environmental advantages.

8.7 Challenges and Future Directions

8.7.1 Limitations and Obstacles in Fungal Remediation

A promising bioremediation method called fungus, commonly referred to as mycoremediation, employs fungi to break down or sequester certain environmental toxins. It has a number of benefits, such as being environmentally friendly and being able to break down complicated contaminants, but it also has significant drawbacks:

1. *Substrate Specificity*: Fungi used in remediation processes often exhibit substrate specificity, meaning they are more efficient at degrading certain types of hydrocarbons or contaminants compared to others. For example, some fungal strains may be highly effective in degrading aliphatic hydrocarbons like paraffin, but less efficient in degrading aromatic hydrocarbons present in asphaltenes. This substrate

specificity can limit the overall effectiveness of fungal remediation when dealing with complex mixtures of hydrocarbons (Alshabib and Onaizi 2019).

2. *Environmental Conditions*: Fungal growth and activity are influenced by various environmental factors, including temperature, pH, moisture content, oxygen availability, and nutrient availability. Suboptimal environmental conditions can limit fungal growth and degrade their remediation efficiency. For instance, low temperatures can slow down fungal metabolic activities, while high temperatures may lead to the inhibition of fungal growth. Additionally, extreme pH levels or inadequate nutrient availability can limit fungal colonization and degradation capabilities.
3. *Slow Degradation Rates*: Fungal degradation processes can be slower compared to other remediation methods, such as chemical or physical treatments. Fungi typically require more time to establish and colonize the contaminated site and degrade the hydrocarbon compounds. The slow degradation rates can be a challenge when there is a need for rapid and efficient cleanup, particularly in cases where time is a critical factor (Daâssi and Almaghrabi 2023).
4. *Scale-up Challenges*: While fungal remediation has shown promise in laboratory-scale studies and small-scale field trials, scaling up the process to larger contaminated sites can present challenges. Factors such as the availability of suitable fungal strains, the logistics of inoculum production, and the uniform distribution of fungi across the contaminated area need to be carefully considered for successful large-scale implementation. Additionally, the cost-effectiveness and feasibility of scaling up fungal remediation projects need to be assessed.
5. *Regulatory Framework*: Fungal remediation techniques may face regulatory challenges and limitations due to the lack of standardized protocols, guidelines, and regulatory frameworks specifically tailored to fungal-based technologies. The absence of clear guidelines or regulatory requirements can hinder the widespread adoption of fungal remediation approaches in some regions. It is important to address regulatory concerns and ensure compliance with relevant environmental regulations during the planning and implementation of fungal remediation projects (Malik et al. 2023).
6. *Monitoring and Validation*: Monitoring the progress and effectiveness of fungal remediation can be challenging. The accurate assessment of fungal growth, activity, and hydrocarbon degradation often requires specialized techniques and analytical methods. Furthermore, validating the complete and long-term removal of contaminants from the site can be complex and time-consuming (Kumar Ravi et al. 2022).

8.7.2 Potential Solutions and Innovations

1. *Strain Selection and Genetic Engineering*: The identification and selection of fungal strains with broad substrate specificity and enhanced degradation capabilities can improve the efficiency of fungal remediation. Research

efforts focus on screening and isolating fungal species that exhibit superior hydrocarbon-degrading abilities. Additionally, genetic engineering techniques can be employed to enhance the metabolic capabilities of selected fungal strains, enabling them to efficiently degrade specific hydrocarbon compounds (Cao et al. 2022).

2. *Co-Cultivation and Microbial Consortia*: Co-cultivation of multiple fungal strains or the use of microbial consortia comprising fungi and other microorganisms can enhance the overall degradation efficiency. Certain fungi may have complementary degradation capabilities, allowing for the synergistic breakdown of complex hydrocarbon mixtures. Co-cultivation or consortia approaches can improve the adaptability of the remediation process to varying environmental conditions and enhance the degradation of a wider range of hydrocarbons (Cao et al. 2022).
3. *Nutrient Optimization and Stimulants*: Optimizing nutrient availability and supplementing with appropriate stimulants can enhance fungal growth and metabolic activity. Nutrient amendments, such as nitrogen, phosphorus, and trace elements, can enhance the growth and activity of fungi. Furthermore, the addition of specific stimulants, such as surfactants or biosurfactants, can improve the bioavailability and accessibility of hydrocarbons, facilitating their degradation by fungi.
4. *Process Optimization and Engineering*: Advancements in process optimization and engineering can improve the scalability, efficiency, and cost-effectiveness of fungal remediation. This includes optimizing parameters such as temperature, pH, moisture content, and oxygen supply to create optimal growth conditions for fungi. Additionally, innovative techniques, such as immobilization of fungal cells, biofilm development, or the use of carrier materials, can enhance fungal colonization and promote sustained degradation in larger-scale applications (Azubuike et al. 2016).
5. *In-situ and Ex-situ Approaches*: *In-situ* fungal remediation involves the direct application of fungi to the contaminated site, while *ex-situ* approaches involve the removal of contaminated materials to a controlled environment for remediation. Advances in both *in-situ* and *ex-situ* techniques, such as bioaugmentation and biostimulation, can enhance the application of fungal remediation in various scenarios. *In-situ* techniques minimize site disturbance, while *ex-situ* approaches offer better control over environmental conditions and allow for the optimization of remediation parameters.
6. *Integration of Fungal Remediation with Other Technologies*: Integration of fungal remediation with other remediation technologies, such as bioreactors, electrokinetics, or phytoremediation, can synergistically enhance the overall efficiency and effectiveness of the remediation process. Combined approaches can target different aspects of contamination, overcome specific limitations, and optimize the removal of paraffin, waxes, and asphaltenes from complex environments (Hassan et al. 2018).

8.8 Conclusion

In conclusion, recyclable environmental-friendly fungi are of paramount importance in sustainable development. Their versatile capabilities in environmental protection, waste management, agriculture, renewable energy, bioremediation, sustainable manufacturing, and climate change mitigation make them valuable assets in promoting a more sustainable and resilient future for our planet. Fungal remediation offers promising solutions for the degradation of paraffin, waxes, and asphaltenes in the oil and gas industries. By understanding the mechanisms involved, optimizing process parameters, and exploring innovative strategies, fungal remediation can be further developed and integrated into existing practices to achieve sustainable and eco-friendly cleanup of hydrocarbon-contaminated environments. The implications for the oil and gas industries in utilizing recyclable, environmentally-friendly fungi for the degradation of paraffin, waxes, and asphaltenes are wide-ranging. From environmental sustainability and cost savings to compliance with regulations and positive public perception, fungal remediation techniques offer numerous benefits for the industry. Embracing these techniques not only addresses hydrocarbon contamination but also contributes to the industry's commitment to environmental stewardship, innovation, and social responsibility.

References

Al-Hawash AB, Alkooranee JT, Zhang X, Ma F (2018) Fungal degradation of polycyclic aromatic hydrocarbons. Int J Pure Appl Biosci 6(2):8–24

Al-Otibi F, Al-Zahrani RM, Marraiki N (2023) Biodegradation of selected hydrocarbons by fusarium species isolated from contaminated soil samples in Riyadh, Saudi Arabia. J Fungi (Basel, Switzerland) 9(2):216. https://doi.org/10.3390/jof9020216

Al-Rashdi A, Al-Hinai FS, Al-Harrasi MMA, Al-Sabahi JN, Al-Badi RS, Al-Mahmooli IH, ... Velazhahan R (2023) The potential of endophytic bacteria from Prosopis cineraria for the control of Pythium aphanidermatum-induced damping-off in cucumber under saline water irrigation. J Plant Pathol 105(1):39–56

Al-Yaari M (2011) Paraffin wax deposition: mitigation & removal techniques. In: SPE Saudi Arabia section young professionals technical symposium, pp SPE-155412. Spe

Almeda R, Wambaugh Z, Chai C, Wang Z, Liu Z, Buskey EJ (2013) Effects of crude oil exposure on bioaccumulation of polycyclic aromatic hydrocarbons and survival of adult and larval stages of gelatinous zooplankton. PLoS ONE 8(10):e74476. https://doi.org/10.1371/journal.pone.0074476

Alshabib M, Onaizi SA (2019) A review on phenolic wastewater remediation using homogeneous and heterogeneous enzymatic processes: current status and potential challenges. Sep Purif Technol 219:186–207

Aranda E (2016) Promising approaches towards biotransformation of polycyclic aromatic hydrocarbons with Ascomycota fungi. Curr Opin Biotechnol 38:1–8

Arun KB, Madhavan A, Tarafdar A, Sirohi R, Anoopkumar AN, Kuriakose LL, Awasthi MK, Binod P, Varjani S, Sindhu R (2023) Filamentous fungi for pharmaceutical compounds degradation in the environment: a sustainable approach. Environ Technol Innov 31:103182, ISSN: 2352-1864, https://doi.org/10.1016/j.eti.2023.103182

Arunkumar N, Gulsar Banu J, Gopalakrishnan N, Prakash AH (2017) Wax degrading bacteria: scope and applications in agriculture. Int J Curr Microbiol Appl Sci 6:649–664

Atlas R, Bragg J (2009) Bioremediation of marine oil spills: when and when not–the Exxon Valdez experience. Microb Biotechnol 2(2):213–221

Azubuike CC, Chikere CB, Okpokwasili GC (2016) Bioremediation techniques-classification based on site of application: principles, advantages, limitations and prospects. World J Microbiol Biotechnol 32(11):180. https://doi.org/10.1007/s11274-016-2137-x

Bala S, Garg D, Thirumalesh BV, Sharma M, Sridhar K, Inbaraj BS, Tripathi M (2022) Recent strategies for bioremediation of emerging pollutants: a review for a green and sustainable environment. Toxics 10(8):484. https://doi.org/10.3390/toxics10080484

Balaji V, Arulazhagan P, Ebenezer P (2014) Enzymatic bioremediation of polyaromatic hydrocarbons by fungal consortia enriched from petroleum contaminated soil and oil seeds. J Environ Biol 35(3):521–529

Benguenab A, Chibani A (2021) Biodegradation of petroleum hydrocarbons by filamentous fungi (Aspergillus ustus and Purpureocillium lilacinum) isolated from used engine oil contaminated soil. Acta Ecol Sin 41(5):416–423

Bianco L, Perrotta G (2015) Methodologies and perspectives of proteomics applied to filamentous fungi: from sample preparation to secretome analysis. Int J Mol Sci 16(3):5803–5829

Boonchan S, Britz ML, Stanley GA (2000) Degradation and mineralization of high-molecular-weight polycyclic aromatic hydrocarbons by defined fungal-bacterial cocultures. Appl Environ Microbiol 66(3):1007–1019

Britannica T (2015) Editors of Encyclopaedia. paraffin hydrocarbon. Encyclopedia Britannica. https://www.britannica.com/science/paraffin-hydrocarbon

Bugg TD, Ahmad M, Hardiman EM, Singh R (2011) The emerging role for bacteria in lignin degradation and bio-product formation. Curr Opin Biotechnol 22(3):394–400

Cao Z, Yan W, Ding M, Yuan Y (2022) Construction of microbial consortia for microbial degradation of complex compounds. Front Bioeng Biotechnol 10:1051233. https://doi.org/10.3389/fbioe.2022.1051233

Chala GT, Sulaiman SA, Japper-Jaafar A (2018) Flow start-up and transportation of waxy crude oil in pipelines-a review. J Nonnewton Fluid Mech 251:69–87

Chandra S, Sharma R, Singh K, Sharma A (2013) Application of bioremediation technology in the environment contaminated with petroleum hydrocarbon. Ann Microbiol 63(2):417–431

Chandra P, Enespa Singh R, Arora PK (2020) Microbial lipases and their industrial applications: a comprehensive review. Microbial Cell Factories 19(1):169.https://doi.org/10.1186/s12934-020-01428-8

Cong S, Tian K, Zhang X, Lu F, Singh S, Prior B, Wang ZX (2019) Synthesis of flavor esters by a novel lipase from Aspergillus niger in a soybean-solvent system. 3 Biotech 9:1–7

Daccò C, Girometta C, Asemoloye MD, Carpani G, Picco AM, Tosi S (2020) Key fungal degradation patterns, enzymes and their applications for the removal of aliphatic hydrocarbons in polluted soils: a review. Int Biodeterior Biodegradation 147:104866

Das N, Chandran P (2011) Microbial degradation of petroleum hydrocarbon contaminants: an overview. Biotechnol Res Int, 941810. https://doi.org/10.4061/2011/941810

Daâssi D, Almaghrabi FQ (2023) Petroleum-degrading fungal isolates for the treatment of soil microcosms. Microorganisms 11:1351. https://doi.org/10.3390/microorganisms11051351

Dell' Anno F, Rastelli E, Sansone C, Brunet C, Ianora A, Dell' Anno A (2021) Bacteria, fungi and microalgae for the bioremediation of marine sediments contaminated by petroleum hydrocarbons in the Omics Era. Microorganisms 9(8):1695. https://doi.org/10.3390/microorganisms9081695

Deshmukh R, Khardenavis AA, Purohit HJ (2016) Diverse metabolic capacities of fungi for bioremediation. Indian J Microbiol 56(3):247–264. https://doi.org/10.1007/s12088-016-0584-6

Dowarah B, Gill SS, Agarwala N (2021) Arbuscular mycorrhizal fungi in conferring tolerance to biotic stresses in plants. J Plant Growth Regul, 1–16

Ekins P, Zenghelis D (2021) The costs and benefits of environmental sustainability. Sustain Sci 16:949–965. https://doi.org/10.1007/s11625-021-00910-5
Elkatory MR, Soliman EA, El Nemr A, Hassaan MA, Ragab S, El-Nemr MA, Pantaleo A (2022) Mitigation and remediation technologies of waxy crude oils' deposition within transportation pipelines: a review. Polymers 14(16):3231. https://doi.org/10.3390/polym14163231
Fallah Shayan N, Mohabbati-Kalejahi N, Alavi S, Zahed MA (2022) Sustainable development goals (SDGs) as a framework for corporate social responsibility (CSR). Sustainability 14(3):1222
Falode O, Olusegun O (2006) Stabilization of some Crude oil emulsions by asphaltenes and reservoir fines. Int J Appl Chem 2:29–43
Fenibo EO, Ijoma GN, Selvarajan R, Chikere CB (2019) Microbial surfactants: the next generation multifunctional biomolecules for applications in the petroleum industry and its associated environmental remediation. Microorganisms 7(11):581
Fraccascia L, Yazdanpanah V, Van Capelleveen G, Yazan DM (2021) Energy-based industrial symbiosis: a literature review for circular energy transition. Environ Dev Sustain 23(4):4791–4825
Freund M, Csikós R, Keszthelyi S, Mózes G (1982) I. Chemical, crystallographical and physical properties of liquid paraffins and paraffin waxes. Dev Pet Sci 14:13–140
Frynas JG (2009) Corporate social responsibility in the oil and gas sector. J World Energy Law Bus 2(3):178–195
Gadd GM (Ed) (2001) Fungi in bioremediation (No. 23). Cambridge University Press
Gadd GM (Ed) (2006) Fungi in biogeochemical cycles (No. 24). Cambridge University Press
Gao H, Zhang J, Lai H, Xue Q (2017) Degradation of asphaltenes by two Pseudomonas aeruginosa strains and their effects on physicochemical properties of crude oil. Int Biodeterior Biodegradation 122:12–22
Gavrilescu M (2010) Environmental biotechnology: achievements, opportunities and challenges. Dyn Biochem Process Biotechnol Mol Biol 4(1):1–36
Ghosh S, Khare SK (2016) Biodegradation of cytotoxic 7-Ketocholesterol by Pseudomonas aeruginosa PseA. Biores Technol 213:44–49
Gorbatiuk LO, Pasichna OO (2020) Toxic impact of oil pollution on fish organism in freshwater and marine ecosystems (a Review). Hydrobiol J 56(6)
Gupta A, Gupta R, Singh RL (2016) Microbes and environment. Principles and applications of environmental biotechnology for a sustainable future, 43–84.https://doi.org/10.1007/978-981-10-1866-4_3
Haider FU, Ejaz M, Cheema SA, Khan MI, Zhao B, Liqun C, Mustafa A (2021) Phytotoxicity of petroleum hydrocarbons: sources, impacts and remediation strategies. Environ Res 197:111031
Hakeem K, Sabir M, Ozturk M, Mermut A (Eds) (2014) Soil remediation and plants: prospects and challenges
Harayama S, Kishira H, Kasai Y, Shutsubo K (1999) Petroleum biodegradation in marine environments. J Mol Microbiol Biotechnol 1(1):63–70
Haritash AK, Kaushik CP (2009) Biodegradation aspects of polycyclic aromatic hydrocarbons (PAHs): a review. J Hazard Mater 169(1–3):1–15
Harzevili FD (2014) Biotechnological applications of the yeast Yarrowia lipolytica, vol 74. Springer, New York, NY, USA
Hassan I, Mohamedelhassan E, Yanful EK, Yuan Z-C (2018) Enhancement of bioremediation and phytoremediation using Electrokinetics. InTech.https://doi.org/10.5772/intechopen.73202
He YM, Duan XG, Liu YS (2014) Enhanced bioremediation of oily sludge using co-culture of specific bacterial and yeast strains. J Chem Technol Biotechnol 89(11):1785–1792
Idris M, Okoro LN (2013) A review on the effects of asphaltenes on petroleum processing. Europ Chem Bull 2(6):393–396
Imam A, Suman SK, Ghosh D, Kanaujia PK (2019) Analytical approaches used in monitoring the bioremediation of hydrocarbons in petroleum-contaminated soil and sludge. TrAC, Trends Anal Chem 118:50–64

Jeon JR, Murugesan K, Baldrian P, Schmidt S, Chang YS (2016) Aerobic bacterial catabolism of persistent organic pollutants—potential impact of biotic and abiotic interaction. Curr Opin Biotechnol 38:71–78

Kadri T, Rouissi T, Brar SK, Cledon M, Sarma S, Verma M (2017) Biodegradation of polycyclic aromatic hydrocarbons (PAHs) by fungal enzymes: a review. J Environ Sci 51:52–74

Kaplan DL (1998) Introduction to biopolymers from renewable resources. In: Biopolymers from renewable resources (pp 1–29). Berlin, Heidelberg: Springer Berlin Heidelberg

Kaskatepe B, Yildiz S (2016) Rhamnolipid biosurfactants produced by Pseudomonas species. Brazilian Arch Biol Technol, 59

Khan AHA, Ayaz M, Arshad M, Yousaf S, Khan MA, Anees M, Iqbal M (2019) Biogeochemical cycle, occurrence and biological treatments of polycyclic aromatic hydrocarbons (PAHs). Iranian J Sci Technol Trans A Sci 43:1393–1410

Khan SR, Nirmal JI, Kumar RN, Patel JG (2015) Biodegradation of kerosene: study of growth optimization and metabolic fate of P. janthinellum SDX7. Braz J Microbiol 46:397–406

Komorowicz M, Janiszewska-Latterini D, Przybylska-Balcerek A, Stuper-Szablewska K (2023) Fungal biotransformation of hazardous organic compounds in wood waste. Molecules 28:4823. https://doi.org/10.3390/molecules28124823

Krüger W, Vielreicher S, Kapitan M, Jacobsen ID, Niemiec MJ (2019) Fungal-bacterial interactions in health and disease. Pathogens (Basel, Switzerland) 8(2):70. https://doi.org/10.3390/pathogens8020070

Kumar Ravi R, Gupta S, Verma R (2022) Bioremediation of petroleum-contaminated soil. IntechOpen.https://doi.org/10.5772/intechopen.100220

Leontaritis KJ (1996) Offshore asphaltene and wax deposition: problems/solutions. World Oil 217(5):57–63

Liu Q, Xia C, Wang L, Tang J (2022) Fingerprint analysis reveals sources of petroleum hydrocarbons in soils of different geographical oilfields of China and its ecological assessment. Sci Rep 12(1):4808

Lv Y, Bao J, Liu D, Gao X, Yu Y, Zhu L (2023) Synergistic effects of rice husk biochar and aerobic composting for heavy oil-contaminated soil remediation and microbial community succession evaluation. J Hazard Mater 448:130929

Machado AL, Lucas EF, González G (2001) Poly (ethylene-co-vinyl acetate)(EVA) as wax inhibitor of a Brazilian crude oil: oil viscosity, pour point and phase behavior of organic solutions. J Petrol Sci Eng 32(2–4):159–165

Maheshwari R, Bharadwaj G, Bhat MK (2000) Thermophilic fungi: their physiology and enzymes. Microbiol Mol Biol Rev 64(3):461–488

Malik S, Bora J, Nag S, Sinha S, Mondal S, Rustagi S, Hazra R, Kumar H, Rajput VD, Minkina T et al (2023) Fungal-based remediation in the treatment of anthropogenic activities and pharmaceutical-pollutant-contaminated wastewater. Water 15:2262. https://doi.org/10.3390/w15122262

Malik A (2008) Bioremediation

Miller ME, Hamann M, Kroon FJ (2020) Bioaccumulation and biomagnification of microplastics in marine organisms: a review and meta-analysis of current data. PLoS ONE 15(10):e0240792. https://doi.org/10.1371/journal.pone.0240792

Mittal A, Singh P (2009) Studies on biodegradation of crude oil by Aspergillus niger. South Pacific J Nat Appl Sci 27(1):57–60

Mohammadi L, Rahdar A, Bazrafshan E, Dahmardeh H, Susan MABH, Kyzas GZ (2020) Petroleum hydrocarbon removal from wastewaters: a review. Processes 8:447. https://doi.org/10.3390/pr8040447

Naeem U, Qazi MA (2020) Leading edges in bioremediation technologies for removal of petroleum hydrocarbons. Environ Sci Pollut Res 27(22):27370–27382

Nagy LG, Riley R, Tritt A, Adam C, Daum C, Floudas D, ... Hibbett DS (2016) Comparative genomics of early-diverging mushroom-forming fungi provides insights into the origins of lignocellulose decay capabilities. Molecul Biol Evolut 33(4):959–970

Nesaratnam ST (1994) A short study on the biodegradation of waste wax from a marine oil terminal. In: Institution of chemical engineers symposium series (Vol 137, pp 179–179). Hemsphere Publishing Corporation

Ngo ACR, Tischler D (2022) Microbial degradation of azo dyes: approaches and prospects for a hazard-free conversion by microorganisms. Int J Environ Res Public Health 19(8):4740

Olajire AA (20212) Review of wax deposition in subsea oil pipeline systems and mitigation technologies in the petroleum industry. Chem Eng J Adv 6, 100104, ISSN 2666–8211, https://doi.org/10.1016/j.ceja.2021.100104

Prince RC, Clark JR (2004) Bioremediation of marine oil spills. In: Studies in surface science and catalysis (Vol 151, pp 495–512). Elsevier

Ragothaman A, Anderson WA (2017) Air quality impacts of petroleum refining and petrochemical industries. Environments 4:66. https://doi.org/10.3390/environments4030066

Ramdass AC, Rampersad SN (2021) Biodiversity and biocatalyst activity of culturable hydrocarbonoclastic fungi isolated from Marac-Moruga mud volcano in South Trinidad. Sci Rep 11(1):19466

Rubin JA, Görres JH (2020) Potential for mycorrhizae-assisted phytoremediation of phosphorus for improved water quality. Int J Environ Res Public Health 18(1):7. https://doi.org/10.3390/ijerph18010007

Saleem M, Moe LA (2014) Multitrophic microbial interactions for eco-and agro-biotechnological processes: theory and practice. Trends Biotechnol 32(10):529–537

Seitmaganbetov N, Rezaei N, Shafiei A (2021) Characterization of crude oils and asphaltenes using the PC-SAFT EoS: a systematic review. Fuel 291:120180

Siles JA, García-Sánchez M (2018) Microbial dynamics during the bioremediation of petroleum hydrocarbon-contaminated soils through biostimulation: an overview. Approaches in bioremediation: the new era of environmental microbiology

Sokal S, Palsania P, Kaushik G (2022) Bioremediation and functional metagenomics: advances, challenges, and opportunities. Omics insights in environmental bioremediation, 3–36

Speight JG (2014) The chemistry and technology of petroleum. 10.b16559

Speight JG (Ed) (1999) The desulfurization of heavy oils and residua. CRC Press

Spina F, Cecchi G, Landinez-Torres A, Pecoraro L, Russo F, Wu B, ... Persiani AM (2018) Fungi as a toolbox for sustainable bioremediation of pesticides in soil and water. Plant Biosyst Int J Dealing Aspects Plant Biol 152(3):474–488

Suryadi H, Judono JJ, Putri MR, Eclessia AD, Ulhaq JM, Agustina DN, Sumiati T (2022) Biodelignification of lignocellulose using ligninolytic enzymes from white-rot fungi. Heliyon 8(2):e08865. https://doi.org/10.1016/j.heliyon.2022.e08865

Tarmizi NFA, Brahmana RK (2023) Environmental performance, political connection, and financial performance: evidence from global oil and gas companies. Environ Sci Pollut Res 30(4):11081–11098

Tiwari N, Garua B, Sharma JG (2021) Microbial diversity, interactions, and biodegradation/biotransformation of organic and inorganic contaminants. In: Wastewater treatment reactors (pp 341–372). Elsevier

Vargas FM, Garcia-Bermudes M, Boggara M, Punnapala S, Abutaqiya M, Mathew N, Al Asafen H (2014) On the development of an enhanced method to predict asphaltene precipitation. In: Offshore technology conference (p D021S017R001). OTC

Vásquez-Murrieta MS, Hernández-Hernández OJ, Cruz-Maya JA, Cancino-Diaz JC, Jan-Roblero J (2016) Approaches for removal of PAHs in soils: Bioaugmentation, biostimulation and bioattenuation. Soil contamination-current consequences and further solutions, 329–342

Wan Ahmad WNK, de Brito MP, Tavasszy LA (2016) Sustainable supply chain management in the oil and gas industry: A review of corporate sustainability reporting practices. Benchmarking Int J 23(6):1423–1444

Wang J, Hao Y, Zhu B, Han T, Li Z, Zhang J (2022) Crystalline behavior of paraffin wax. J Phys Chem B 126(4):985–995

Xenia ME, Refugio RV (2016) Microorganisms metabolism during bioremediation of oil contaminated soils. J Biorem Biodegrad 7(2):20–25
Yang Z, Zhang Z (2018) Engineering strategies for enhanced production of protein and bio-products in Pichia pastoris: a review. Biotechnol Adv 36(1):182–195
Zhang T, Zhang H (2022) Microbial consortia are needed to degrade soil pollutants. Microorganisms 10(2):261
Zheng F, Shi Q, Vallverdu GS, Giusti P, Bouyssiere B (2020a) Fractionation and characterization of petroleum asphaltene: focus on metalopetroleomics. Processes 8:1504. https://doi.org/10.3390/pr8111504
Zheng F, Shi Q, Vallverdu GS, Giusti P, Bouyssiere B (2020b) Fractionation and characterization of petroleum asphaltene: focus on metalopetroleomics. Processes 8(11):1504

Chapter 9
Use of Envrionmentally-Safe Microalgae for the Degradation of Organic Deposit in Crude Oil

Jutishna Bora, Sagar Mondal, Sumira Malik, Smita Lata, Nayan Talukdar, and Sarvesh Rustagi

Abstract The need for technical and industrial transformation has grown highly reliant on the accessibility and utilization of petroleum-based goods as an energy source for socio-economic progress. However, it is important to note that petrochemical and petroleum-based goods have a significant connection to worldwide economic operations. The widespread distribution, refining procedures, and ultimate pathways into the environment of these products present a potential danger to both human health and the ecology. Further worldwide environmental concerns arise from the toxicological effects of hydrocarbon pollutants on air, soil, and water, which have carcinogenic properties for both animals and humans. Hence, it is pragmatic to employ biodegradation using a biological catalyst to tackle the restoration of petroleum-contaminated ecosystems, mitigate unfavorable effects, deal with the intricate nature of hydrocarbons, and overcome resistance to biodegradation. This review explores the use of living organisms to eliminate petroleum hydrocarbon contaminants from water and soil, with a particular emphasis on microorganisms capable of degrading petroleum pollutants. It also examines the processes of mineralization and the transformation of complex organic and inorganic pollutants into simpler compounds through biological agents. Additionally, the review discusses the impact of physicochemical and biological factors on biodegradation mechanisms

J. Bora (✉) · S. Mondal · S. Malik · S. Lata
Department of Biotechnology, Amity University Jharkhand, Ranchi, Jharkhand, India
e-mail: jbora@rnc.amity.edu

S. Mondal
e-mail: sagar.mondal@s.amity.edu

N. Talukdar
Program of Science, Faculty of Biotechnology, Assam down town University, Guwahati, Assam, India

S. Rustagi
School of Applied and Life Sciences, Uttaranchal University, Dehradun, Uttarakhand, India
e-mail: sarveshrustagi@uttaranchaluniversity.ac.in

S. Banerjee et al. (eds.), *Environmental Friendly Green Technologies for Improvement of Heavy Crude Oil Flow Assurance*, Environmental Science and Engineering,
https://doi.org/10.1007/978-3-031-86701-9_9

and enzymatic systems. Recent research on bioremediation strategies for addressing petroleum spills, highlighting their commercial potential, is also featured.

Keywords Bioremediation · Microalgae · Biodegradation · Pollutants · Hydrocarbons

9.1 Introduction

The term "petroleum" originated from the combination of the words "petra" (meaning stone) and "oleum" (meaning oil). Petroleum is a naturally occurring substance that is often refined into different products by a refinery. Crude oil is a type of petroleum that consists mainly of hydrocarbons, along with certain heteroatom compounds and minor amounts of metallic elements. The aliphatics, aromatic compounds, asphaltenes, as well as resins are broad categories that can be used to classify different types of petroleum derivatives. The aliphatic fraction comprises of linear alkanes, alkanes with branches (iso-alkanes), as well as cyclic alkanes (naphthenes). The aromatic fraction consists of volatile monoaromatic hydrocarbons, including benzene, toluene, and xylenes, as well as polycyclic aromatic hydrocarbons (PAH), naphthenic aromatics, and aromatic sulphur compounds such as thiophenes as well as dibenzothiophenes. The asphaltene component consists of polar molecules containing ketones, phenols, fatty acids, esters, along with porphyrins, while the resin component is made up of polar molecules containing sulfoxides, amides, pyridines, quinolines, as well as carbazoles. These polar molecules in both components contain nitrogen (N), sulfur (S), and oxygen (O2). The distribution of these fractions is determined by various factors such as the source, age, past geologic migration, and various other relevant factors of the crude oil (Darmayati 2009). Although petroleum molecules are intricate, they can be generically categorized into two primary groups: hydrocarbons (estimated as THP) as well as non-hydrocarbons. Crude oil is being fractionated through various methods, but it remains predominantly an intricate blend of chemicals. Jet fuels consist of almost 300 distinct hydrocarbon molecules. Spills are typically the most serious kind of pollutants that contaminate the environment, including soil, water, and oceans. This review focusses mostly on water and soil. Leakage and regular washing have been recognized as prevalent sources of contamination. Additional sources of oil spills include subterranean storage tanks, offshore structures, drilling operations, maritime incidents, damaged pipelines, and naturally occurring oil spills.

9.2 Organic Pollutants Crude Oil Derivatives

9.2.1 *Organic Dyes*

Chemical and electrochemical techniques, such as adsorption as well as ultrafiltration, are ineffective in eliminating organic dyes from water due to their limited decomposability. Consequently, organic dyes comprise a significant source of pollution in water. These contaminants also harm aquatic photosynthesis, which serves as the foundation of the food chain. Anthraquinone, azo, along with phthalocyanine represent the three primary classifications of dyes, determined on the chemical compositions that contain their chromophore categories. Cationic basic dyes can be classified into two groups: dyes that do not disperse in an ionic solution, as well as dyes which can be anionic-direct, acidic in nature or even reactive organically. The adsorption approach is a cost-effective and convenient technology for the decolonisation process, making it frequently used (Ali and Muhammad 2008).

Azo and anthraquinone dyes, which come in a wide range of colors, are the most commonly used dyes worldwide. There are over 3000 different varieties of these dyes. They are present in various manufacturing sectors, including as textiles, plastics, as well as medicine. An abundance of effluents with high color content, which exhibit toxicity at a relatively low level, is held responsible for detrimental impacts. The dyes are also highly accumulated, leading to eutrophication and a restricted ability for reoxygenation. Among the most significant issues is the production of toxic amines when azo dyes decompose. Consequently, the majority of aquatic creatures perish as a result of their toxicity.

9.2.2 *Organic Hydrocarbons*

The main categories that constitute organic hydrocarbon pollutants are polycyclic aromatic hydrocarbons (PAHs), polychlorinated biphenyls (PCBs), di-(2-Ethylhexyl) phthalate (DEHP), surfactants, as well as polychlorinated dibenzo-p-dioxins and dibenzofurans (PCDD/PCDF). These chemicals are hydrophobic and have the ability to persist in water bodies for an extended period of time. They are commonly found in association with sediments. Petroleum goods are the main cause of pollution from hydrocarbons in the water environment. These compounds are commonly introduced into marine ecosystems through rivers, coastal waters, industrial activities, and the atmosphere as a result of natural processes that occur during the earth's biogeochemical cycles (Khan et al. 2019). In addition, these compounds exhibit gentle decomposability and resistance to degradation in aerobic conditions in aquatic environments. The rate of degradation decreases significantly as the quantity of aromatic chains as well as molecular weight rise. PAHs, or polycyclic aromatic hydrocarbons, are categorized as carcinogens. They pose a significant risk to the environment because of their toxicity and their ability to persist for a long time without breaking down. Personal care products (PCPs) exert multiple adverse consequences on aquatic creatures, such as disrupting endocrine, developmental, as well as epigenetic systems, which as well as affecting human health. Phenol, a significant carcinogenic organic pollutant, is commonly present in the waste discharge of various industries, such as petroleum refineries. Its concentration levels typically vary between 13 and 88 parts per million (ppm). The levels of 5–25 ppm pose a danger to marine life. Conventional methods are not successful in eliminating phenol compounds from water systems. Nevertheless, biological approaches show promise as a result of the organisms' great efficiency in removing contaminants.

9.3 Biodegradable Pollutants

Recently, highly poisonous chemical compounds were deliberately or inadvertently produced and discharged into the nearby environment. The compounds encompass insecticides, coloring agents, polychlorinated biphenyls (PCB), fuels, along with polycyclic aromatic hydrocarbons (PAH). Synthetic chemicals, which include radionuclides and metals, have a strong resistance to biodegradation by natural flora, in contrast to naturally existing organic molecules, which readily disintegrate when introduced to the environment.

9.3.1 Hydrocarbons

These hydrocarbons can be observed in the form of either aromatic or aliphatic compounds, which can exist as linear, branched, or cyclic structures. The previous isolation consisted of benzene (C6H6) in its composition, whereas the aliphatic form was found in three variations: alkanes, alkenes, or alkynes.

9.3.2 Polycyclic Aromatic Hydrocarbons (PAHs)

Hydrophobic organic pollutants (HOCs), which are important pollutant classes, are widely found in the air, soil, and sediments. The principal cause of PAH pollution is industrial production. PAHs are found in soils and sediments that have a high concentration of organic matter, leading to their accumulation in fish as well as other aquatic creatures. The most probable mode of transmission to humans is through the consumption of seafood. The biodegradation of PAHs can be considered as a conventional process within the carbon cycle and as a release of synthetic pollutants from the ecosystem. Nevertheless, the utilization of microorganisms for the restoration of areas contaminated with polycyclic aromatic hydrocarbons (PAH) seems to be a highly promising approach for the purification of polluted sites.

9.3.3 *Polychlorinated Biphenyls (PCBs)*

These are artificially created chemical compounds. Due to their non-flammable nature, chemical stability, high boiling point, as well as electrical insulating properties. PCBs have been utilized for numerous industrial applications. The items encompass electrical, heat transfer, along with hydraulic components, as well as solvents found in paint, plastic, even rubber goods. Additionally, pigments, dyes, as well as carbonless copy paper are also included. The increasing problem of environmental contamination caused by PCBs is due to their potential as endocrine disruptors and carcinogens (Joutey et al. 2013).

9.3.4 *Pesticides*

These compounds or their combination are designed to prevent, exterminate, deter, or diminish pests. Non-biodegradable chemicals undergo rapid degradation, while persistent molecules impede disintegration. The predominant form of breakdown in soil is caused by microorganisms, particularly fungus and bacteria, which metabolize pesticides.

9.3.5 *Dyes*

Dyes are widely employed in various industries such as textiles, rubber manufacturing, paper production, printing, photography, medicines, cosmetics, and several other sectors. Azo dyes, which are widely used in commercial applications, are intrinsically resistant to biodegradation due to their chemical structures. In general, the treatment of wastewater containing dyes usually involves many physical and chemical methods, including adsorption, coagulation-flocculation, oxidation, purification, including electrochemical processes, which have an impact on the processes (Joutey et al. 2013).

9.3.6 *Radionuclides*

The nuclei of these atoms are unstable, meaning they have excess energy that can be transferred either to a newly formed particle inside the nucleus or by a process called internal conversion. Radionuclides undergo radioactive decay, resulting in the release of gamma rays along with subatomic particles that include alpha or beta particles (Lloyd and Lovley 2001).

9.4 Types of Hydrocarbons Degradation

The enzymatic reactions involved during the aerobic and anaerobic breakdown of hydrocarbons by bacteria have been extensively studied for many decades. This section focusses on the degradation of aliphatic, aromatic, resin, along with asphaltene compounds.

9.4.1 Aliphatic Degradation

Aliphatics, specifically alkanes, are a major component of crude oil. The removal of these compounds from oil-contaminated areas is a matter of ecological concern but is believed to be advantageous for enhancing oil recovery (Hidayat and Turjaman 2016). The degradation of alkanes by microorganisms can occur through many mechanisms. The mechanisms involve the formation of alkyl hydroperoxide, terminal oxidation, as well as subterminal oxidation, which undergo enzymatic and non-enzymatic reactions. Aliphatic degradation is often categorized according to the following ascending levels of vulnerability: N-alkanes are higher in priority than branched alkanes, while branched alkanes are higher in priority than cyclic alkanes.

9.4.2 Aromatic Degradation

This type of compound contains a couple of aromatic rings, with benzene being the most basic example. The aromatic compounds found in crude oil, whether they include alkyl substituents or not, along with their fractions, are regarded as the second most significant category in crude oil, following the aliphatic fraction. Compounds containing two or more benzene rings that are joined together are categorized to be polycyclic aromatic hydrocarbons (PAHs). PAHs are a significant class of foreign substances that are widespread, long-lasting, and resistant to pollutants. According to reports, several of them are extremely mutagenic or carcinogenic, indicating their potential hazard.

9.4.3 Resin and Asphaltene Degradation

Both of these the resin along with asphaltene fractions consist of polar non-hydrocarbon compounds, which is different from the aliphatic as well as aromatic fractions. The former mostly consists of hydrogen as well as carbon, with little amounts of nitrogen, sulfur, and oxygen. Asphaltenes are complex compounds with a large molecular size that cannot dissolve in solvents such as n-heptane. Resins have

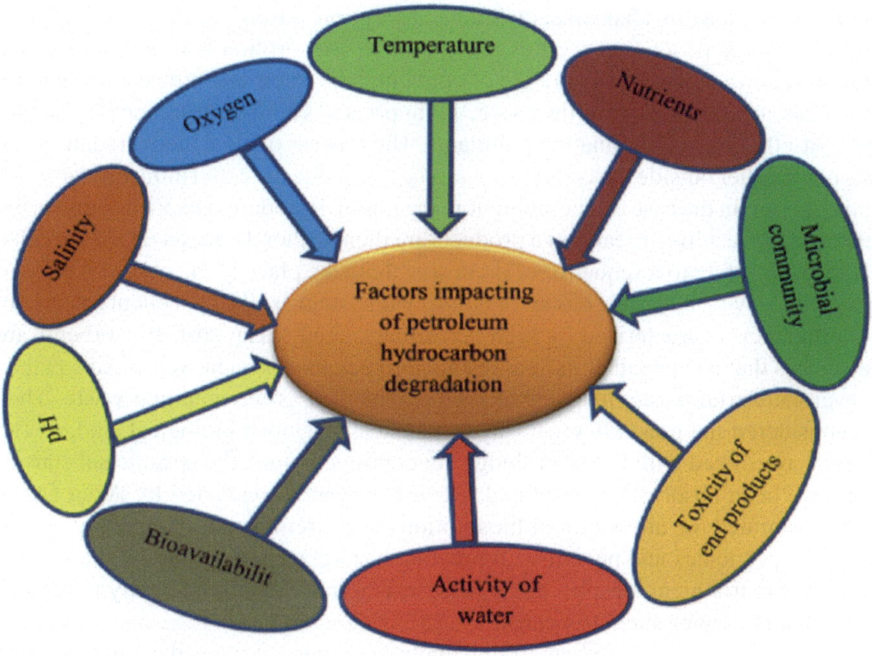

Fig. 9.1 Schematic diagram of factors affecting Petroleum hydrocarbon degradation

polar molecules which can be dissolved in n-heptane, while asphaltenes are complex structures that provide challenges in terms of comprehension and enhancement through biodegradation (Rahayu 2020).

The factors that affect petroleum hydrocarbon degradation is represented in Fig. 9.1.

9.5 Algae as a Natural Biodegradation Agent

Algae have a significant role in the natural method of treating wastewater, known as "phycoremediation" (Fig. 9.2). Algae have the ability to gather and store harmful organic and inorganic toxins, as well as radioactive chemicals, within their cells. As a result, they have a significant role in the process of purifying municipal, industrial, and agro-industrial wastewater. Some of them are shown in Table 9.1. Additionally, they play a significant role in water systems as the primary producer. Various types of organic waste can pollute water, leading to the growth of multiple algae species. These algae have developed several mechanisms to detoxify the polluted water, such as biosorption, bioaccumulation, biotransformation, biomineralization, and in situ and/or ex situ biodegradation. Biosorption is a physicochemical process that relies on many mechanisms, including absorption, adsorption, surface complexation, exchange of ions, as well as precipitation. Biosorption is a highly effective method for eliminating a wide range of contaminants, such as textile dyes, phenolic chemicals, and pesticides. Furthermore, this approach is both environmentally benign and cost-effective for eliminating pollutants. The process of algal biodegradation can take place either outside the cells (extracellularly), inside the cells (intracellularly), or a mix of both. In the case of a combination, the initial degradation happens outside the cells, and the resulting breakdown products are then further degraded inside the cells. Biosorption refers to a sequence of processes that take place in the cell wall prior to anaerobic as well as aerobic biodegradation. The system exhibits excellent selectivity and efficiency, characterized by high performance and cheap cost. Biosorbents are substances that can be either natural resources, such marine algae or invasive plants, or even industrial waste, like activated sludge as well as fermentation waste. They are considered the most effective way to physically remove biological sludge. The hazards associated with biomass sludge are contingent upon the specific substances that have been absorbed or absorbed. Review research conducted by Aksu (Aksu 2005) examined the utilization of biosorption for the removal of organic pollutants, including pesticides and phenols. Eliminating hazardous and noxious contaminants from the environment, despite the high cost, is a more environmentally conscious option than allowing them to persist.

Biosorption is considered environmentally benign since it has the ability to both absorb and breakdown contaminants, including azo dyes, producing simpler chemicals. Algae and cyanobacteria have the ability to break down pollutants by releasing

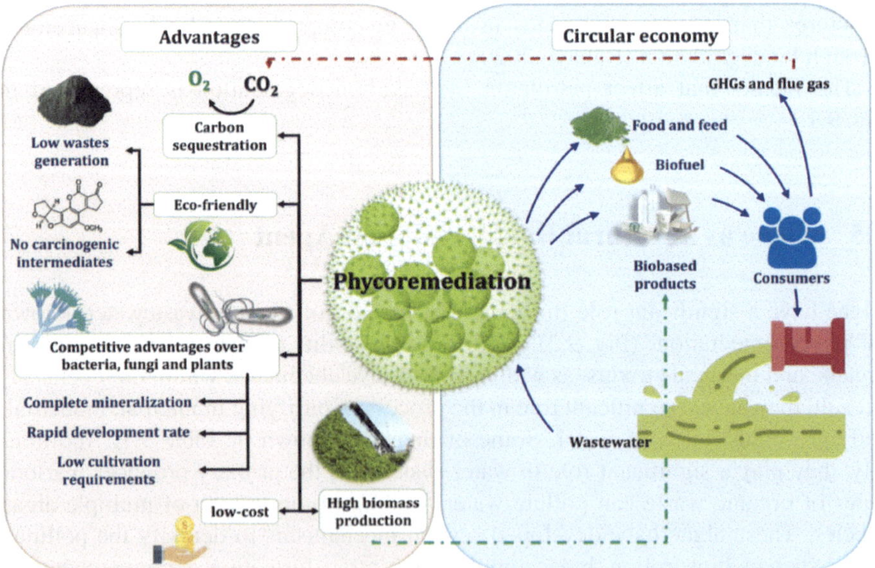

Fig. 9.2 The advantages and effective roles of algae in phycoremediation of wastewater

Table 9.1 Eukaryotic microalgae capable of breaking down petroleum hydrocarbons

Genera	Substrata	References
Phylum Chlorophyta		
Ankistrodesmus capricornutum	Benzo(a)pyrene	Paniagua-Michel and Banat (2024)
Chlamydomonas reinhardii	Phenanthrene	
Chlorella vulgaris	Naphthalene Nonadecane	
Dunaliella salina	Naphthalene	
Prototheca zopfii	n-alkanes and PAHs	
Scenedesmus acutus	Benzo(a)pyrene	
Phylum Heterokontophyta		
Class Bacillariophyta		
Achnanthes minutissima	Alkanes	Prince (2010)
Cyclotella caspia	Fluoranthene	
Cylindrotheca sp.	Naphthalene	
Diatom BD1IITG	Phenol	
Nitzschia sp., Navicula sp.	Naphthenic acid	
Nitzschia sp.	Phenanthrene	
Navicula sp.	Naphthalene	
Navicula sp., Nitzschia sp.,	Naphthalene	
Synedra sp., Amphora sp.	Naphthalene	
Skeletonema costatum	Fluoranthene	
Thalassiosira sp.	Phenol	
Nitzchia sp. and Bacteria	Benzo(a)pyrene	
Genera Marivita, Erythrobacter, Alcaligenes	Fluoranthene	
Cylindrotheca sp., Amphora sp.	Naphthalene	

specialized enzymes, like azo-dye reductase, which can transform dyes into harmless substances such as NH2 and CO2. There is a lack of study on the effectiveness of phycoremediation in decreasing nutrient amounts in eutrophic lakes. However, Razak and Sharip (Rahayu 2020) explored the potential of using algae-based methods to address the issue of lake eutrophication. Most phycoremediation research primarily focus on wastewater, according to their findings. Employing microalgae to diminish fertilizer, pollution, and heavy-metal concentrations in lakes is an effective method to enhance water quality. Researchers have given scant attention to lake phycoremediation. Hence, it is imperative to conduct further research on the ecological as well as physiological characteristics of algae, particularly in relation to eutrophication. Phycoremediation's efficacy should be assessed in authentic environments, such as ponds and lakes. Bioremediation employs microorganisms, including bacteria, fungi, and algae, which possess the ability to enzymatically degrade harmful substances.

Nutrient enrichment stimulates the proliferation of indigenous algae in all aquatic environments. These native bacteria function in a similar manner to those employed in bioremediation techniques, while algae release oxygen around the atmosphere. Photosynthetic algae contribute to the production of oxygen in the atmosphere, hence enhancing water quality, aquatic life, including biodiversity. Phycoremediation is the most efficient way for removing nutrients from lakes, as it helps preserve the lakes' natural food chain. Algae and indigenous microorganisms collaborate to purify water. Algae utilize carbon dioxide (CO2) and nutrients to produce oxygen (O2), and there is no production of waste in either mixotrophic or heterotrophic circumstances.

Phycoremediation is a secure solution as algae efficiently transform sunlight into valuable biomass by utilizing pollutants like phosphorus and nitrogen, which are the main causes of eutrophication. Both viable and nonviable algae have been demonstrated to be effective in speeding up the process of biodegradation and transforming organic contaminants into simpler compounds. Several research have examined the utilization of microalgae and cyanobacteria populations for the process of phytoremediation, specifically targeting the removal of organic contaminants in water bodies.

Cyanobacterial species are known for their exceptional capacity to eliminate organic contaminants from water bodies due to their robust photosynthetic activity, leading to the production of large biomasses. These biomasses include a diverse range of bioactive compounds that can be utilized in various cost-effective applications.

Cyanobacteria as well as eukaryotic microalgae can be differentiated by their ability to develop quickly in aseptic environments. Algal biomass, which contains a significant amount of polyunsaturated fatty acids, has been previously utilized for several applications including biodiesel, biogas, as well as bioethanol production. The rate at which pollutants are eliminated and broken down depends on their quantity, molecular weight, biomass from algae, species, the metabolic rate of algae, development phase, and ambient variables. Several algae belonging to different classes flourish and develop most effectively in polluted conditions, such as *Euglena viridis* (Euglenophyta), *Nitzschia palea* (Bacillariophyta), *Oscillatoria limosa, O. tenuis, O. princeps*, as well as *Phormidium uncinatum* (Cyanophyta).

Algae employ three distinct mechanisms to eliminate organic contaminants, such as PAHs, from ecosystems. One way is through the adsorption of PAHs onto the algal cell wall, where active groups attract and form complexes through electrostatic attraction. The second mechanism involves the bioaccumulation of contaminants within cellular structures. The biosorption and aggregation ability of algae in this process is linked with their functional groups, such as hydroxyl, carboxylate, sulphate, phosphate, as well as amino groups (Khan et al. 2020).

Concurrently, the biosorption process encompasses a range of chemical reactions taking place on the surface of the cell. The reactions involved in this process include ion exchange with the active groups present on the surface of the algae, surface complexation reactions, the chelation, as well as microprecipitation. One method to eliminate organic pollutants is by transforming them into useful substances, such as water carbon dioxide, by the use of redox enzymes that participate in enzymatic reactions involving oxidation and reduction. Microalgae possess numerous enzymes that aid in the degradation of diverse chemical substances that induce cellular stress within microalgae. These enzymes are complex and are involved in both phase I along with phase II processes. Monooxygenases, dioxygenases, hydroxylases, carboxylases, as well as decarboxylases are crucial in the process of biodegradation carried out by phase I enzymes. They contribute to the enhancement of the hydrophilicity of pollutants by introducing or revealing a hydroxyl group through oxidation, reduction, or hydrolysis reactions. Phase II involves the action of an enzyme that facilitates the combination of glutathione as well as glutathione-S-transferases with various substances that have electrophilic sites. This process leads to the breaking of the epoxide ring and provides defense against oxidative stress within the cell. Laccase glycoproteins are enzymes that are located outside of cells and have a crucial function in the breakdown of various organic contaminants by microalgae. They either exhaust the processes or undergo cross linking afterwards. Another theory proposes that the mechanism of algal biodegradation could replace standard water treatment technologies and be a highly effective tool for removing various pollutants. Biosorption occurs when contaminants in a water-based solution are either adsorbed onto algal cell-wall components or onto organic compounds ejected by algal cells, particularly protein as well as extracellular polysaccharides that are released into the surrounding environment. This approach is passive and relies on various elements including the kind of biosorbents, ambient conditions, and algae metabolic processes. Bioaccumulation occurs when organic contaminants are transferred into a food chain through algae, which is influenced by algal metabolic activity. Eutrophication is the predominant adverse consequence of pollution. Excessive quantities of organic matter and decomposing organisms lead to oxygen depletion in water, resulting in the death of other species. The nutrients NH^{4+}, $NO3^-$, and $PO43^-$ found in secondary effluents represent the main cause of eutrophication within natural bodies of water. These nutrients contribute to the deterioration of water quality and have negative effects on aquatic ecosystems. Consequently, prior to being discharged into bodies of water, wastewater must undergo proper treatment. Several unit processes, such as activated sludge (for the purpose of phosphorous and nitrogen removal), electro flocculation, filtration via membranes, electrokinetic coagulation, electrochemical annihilation exchange of ions, CatOx treatment (catalytic oxidation), as well as disinfection (by ozonation, chlorination, along with ultraviolet light), are employed to eliminate nutrients from wastewater (Aksu 2005).

Algae secrete different enzymes that are important in breaking down various organic contaminants into less harmful substances. In addition, they have catabolic genes that enable them to break down various contaminants. Algae that produce biosurfactants can greatly improve the process of breaking down hydrocarbon pollutants in contaminated areas. The biosurfactants consist of various intricate biomolecules such as glycolipids, lipoproteins, lipopeptides, lipids that are neutral, phospholipids, as well as fatty acids. These biosurfactants have amphiphiles, which means they can increase the solubility of hydrophobic contaminants in water. This helps to speed up the process of hydrocarbon bioremediation. The biodegradation of PCBs by algae is influenced by the physicochemical as well as physiological properties of PCBs, such as their concentration and hydrophobicity. It also depends on the biomass as well as physiological capabilities of the algae, as well as their membrane permeability (Stange and Swackhamer 1994).

9.6 Bioremediation of Organic Pollutants by Microalgae

Phycoremediation, using microalgae species, is a highly successful method for promoting environmental sustainability. Microalgae has a high surface-area-to-volume ratio, rapid metabolism, and are readily available, making them excellent at breaking down various environmental pollutants. Additionally, this approach is cost-effective. Microalgae undergo biological amplification, wherein they absorb and accumulate various poisons, which can then be transferred to higher levels of the food chain (Correa-Reyes et al. 2007). The ability of algae to accumulate substances is contingent upon the conditions in which they grow and is influenced by their lipid content. Several taxa, such as *Botryococcus, Chlamydomonas, Chlorella, Scenedesmus, Nitzschia,* as well as *Micractinium,* are suggested for the treatment of wastewater. Moreover, these species exhibit a substantial oil yield, which might be utilized for the production of biodiesel.

9.7 Conclusions and Future Perspectives

The widespread utilization of hydrocarbon waste along with other organic pollutants across various systems has led to their emergence as one of the most significant environmental issues worldwide. Microalgae and cyanobacteria offer a green and sustainable solution for treating polluted water by removing different types of organic contaminants. This bioremediation method has a lower environmental impact compared to other microorganisms as well as traditional approaches. Moreover, these pollutants facilitate the proliferation of algal biomass, that can be harnessed for various purposes in the future. In order to address this challenge, future efforts would be focus on screening a wider range of algae strains towards phycoremediation, and possibly utilizing genetic engineering techniques to enhance the ability of algae to degrade organic contaminants and increase their tolerance to different types of pollutants.

Furthermore, the transfer of newly discovered genes from bacteria that can break down pollutants would be used to speed up the rate at which a wide range of organic contaminants are broken down in algae. In order to expedite the process of using living organisms to clean up organic pollutants and decrease the time it takes to decontaminate, it is necessary to modify the physical and chemical properties of aquatic environments, including temperature, pH, as well as nutrient levels. Efficient design of the expanding system is crucial for optimizing expansion rates and reducing expenses.

References

Aksu Z (2005) Application of biosorption for the removal of organic pollutants: a review. Process Biochem 40(3–4):997–1026

Ali H, Muhammad SK (2008) Biosorption of crystal violet from water on leaf biomass of Calotropis procera. J Environ Sci Technol 1(3):143–150

Correa-Reyes G et al (2007) Nonylphenol algal bioaccumulation and its effect through the trophic chain. Chemosphere 68(4):662–670

Darmayati Y (2009) Development of oil bioremediation research on marine environment in Indonesia. Development 12:105–110

Hidayat A, Turjaman M (2016) Biological degradation of crude oil contaminants and its Application in Indonesia. Bagyaraj, DJ

Joutey NT, Bahafid W, Sayel H, El Ghachtouli N (2013) Biodegradation: involved microorganisms and genetically engineered microorganisms. Biodegradation-Life of Science 1:289–320

Khan AHA et al (2019) Biogeochemical cycle, occurrence and biological treatments of polycyclic aromatic hydrocarbons (PAHs). Iranian J Sci Technol Trans A Sci 43:1393–1410

Khan ZI et al (2020) A study on the transfer of chromium from meadows to grazing livestock: an assessment of health risk. Environ Sci Pollut Res 27:26694–26701

Lloyd JR, Lovley DR (2001) Microbial detoxification of metals and radionuclides. Curr Opin Biotechnol 12(3):248–253

Paniagua-Michel J, Banat IM (2024) Unravelling diatoms' potential for the bioremediation of oil hydrocarbons in marine environments. Clean Technol 6(1):93–115

Prince R (2010) 16 Bioremediation of marine oil spills

Rahayu YS (2020) Bioremediation model of oil-contaminated soil in Lapindo mud using multisymbiotic organism. Manage Environ Qual Int J 31(3):586–601

Stange K, Swackhamer DL (1994) Factors affecting phytoplankton species-specific differences in accumulation of 40 polychlorinated biphenyls (PCBs). Environ Toxicol Chem Int J 13(11):1849–1860

Chapter 10
Microbial Remediation of Petroleum Hydrocarbon Polluted Soils: Challenges and Perspectives

Sayanti De, Soumyadeep Biswas, Subhra Satahrada, Nirmalya Pal, Sandeep Kumar Panda, and Ritesh Pattnaik

Abstract As a consequence of rapid industrialization and urbanization phenomenon, extensive load and impact on petroleum-based industries occur towards the rising demands. In a global perspective, the development of petroleum industries involves both extraction and purification processes. However, accidental oil leaks and spills can often result in severe soil and groundwater pollution, leading to drastic change in physical, chemical and biological properties in the environment. Therefore, to manage and mitigate pollution and environment toxicity, the petroleum-contaminated soil (PCS) either due to oil spillage or release require efficient and cost-effective remediation. Worldwide researchers are continuously investigating and providing effective strategies towards development of various remediation approaches to repair and re-control the afflicted environment due to petroleum hydrocarbon contamination. Over the past years, bioremediation strategies have emerged as an eco-friendly and cost–effective alternative in comparison with the traditional physico-chemical approaches for the restoration and reclamation of contaminated sites. The present chapter outline the diversified role and capacities of microorganisms that can potentially degrade petroleum hydrocarbons in polluted soils. In addition, insights are provided about the metabolic genes, various pathways of microbial degradation of petroleum pollutants, and the internal–external aspects, and challenges that influence remediation in order to select the optimal remediation treatment strategy.

Keywords Petroleum hydrocarbons · Polluted soils · Contamination · Bioremediation · Metabolic pathways

---

S. De · S. Biswas · S. Satahrada · N. Pal · S. K. Panda (✉) · R. Pattnaik (✉)
School of Biotechnology, Kalinga Institute of Industrial Technology (KIIT) Deemed to be University, Bhubaneswar 751024,, Odisha, India
e-mail: sandeep.panda@kiitbiotech.ac.in

R. Pattnaik
e-mail: rpattnaik@kiitbiotech.ac.in

S. Banerjee et al. (eds.), *Environmental Friendly Green Technologies for Improvement of Heavy Crude Oil Flow Assurance*, Environmental Science and Engineering,
https://doi.org/10.1007/978-3-031-86701-9_10

10.1 Introduction

Petroleum Hydrocarbons (PHCs) are complex compounds formed from hydrogen and carbon molecules. PHCs are primarily made up of alicyclic or aromatic compounds and short chain hydrocarbons along with occurrence of various other contaminants like oxygen, nitrogen or sulphur compounds. Petroleum hydrocarbons are widely used as primary source of energy for household and industrial activities. With the rapid industrialization and urbanization phenomenon, there is a great demand for petrochemical products. During the process of fulfilling the demands, PHCs are frequently contaminated during exploration, production, maintenance, transport, storage, and unintentional accidents. The formation of petroleum takes place via two steps (a) Diagenesis and (b) Catagenesis.

Diagenesis: Proteins, carbohydrates, lipids and other such organic sediments, rich in minerals, are saturated with water and buried deep in the soil. A series of chemical, microbial and compaction processes take place which extract the water out of these. The proteins and carbohydrates further break down to give rise to Kerogen, a waxy substance and Bitumen, a black tar substance. This process can happen under mild conditions of temperature and pressure.

Catagenesis: Also referred to as "cracking", this process turns kerogen into petroleum and natural gas. When the temperature and pressure increases (deeper burial) kerogen start to undergo thermal degradation to form hydrocarbon chains. The mineral deposited in the soil help to catalyse this reaction. The cracking process works better in conditions of higher temperature and pressure and thus smaller and lighter hydrocarbons are formed.

Four classes of PHCs are present which includes:

(a) Aliphatic Compounds: They comprise of n-alkanes, iso-alkanes, cycloalkanes, etc. Examples: Methane, Ethane, Butane (n-alkanes), Isobutane, Neopentane, Isooctane (iso-alkanes)
(b) Aromatic Compounds: They are monocyclic, polycyclic and chlorinated aromatic compounds. Example: Benzene, Toluene, Xylene, Cumene, etc.
(c) Resins: They contain aromatic compounds with long alkyl chains.
(d) Asphaltenes: These compounds have a large molecular weight and are viscous. They are made up of polycyclic clusters that have different alkyl groups substituted throughout.

10.2 Effect of Petroleum Hydrocarbon on Environment

Petroleum (also known as crude oil) is essentially has carbon, hydrogen of 85% and 15% respectively, as well as a compound mixture of hydrocarbons such paraffins, naphthene, aromatic hydrocarbons, and vaporous hydrocarbons (ranging from CH_4 to C_4H_{10}). In addition to these, crude oil also contains small amounts of minerals and non-hydrocarbons (such as sulphur, nitrogen, and oxygen mixes); the sulphur content of heavier crudes is higher. Depending on the strength of the hydrocarbons, petroleum can be classed as paraffin base, ordinary base, or naphthenic base. PHCs contamination of the soil is a worldwide concern. Oil and fuel spills in the soil are among the major reasons of soil contamination. Refinery products break down the biochemical and physiochemical characteristics of the soil, which limits plant growth and development. It causes a deficiency of water, oxygen, nitrogen and phosphorous which are important for the plants to develop like root development reduces, reduction in height, no. of leaves, chlorophyll content in plants and hindered germination processes. The petroleum-contaminated soils also poison the groundwater, limiting its usage, resulting in financial loss, ecological problems, and a reduction in the soil's ability to support agricultural activities. The PHCs discharged into the ocean, amid transportation, promotes contamination due to low dissolvability of crude oil in water. It leads to agglomeration of particles which leads to poor oxygen exchange. Other heavier portions collect at the base of the water which makes the survival of the aquatic organisms critical.

10.2.1 Conventional Remediation Approaches for PCS

10.2.1.1 Physical Remediation Approaches

Due to the serious environmental problems, many techniques of remediation have evolved throughout time. The traditional in-situ and ex-situ remediation methods are listed here. The strategies can be divided into physical, chemical and biological remediation and are as follows:

(A) Soil Vapour Extraction: It is an in-situ approach in which extraction well is employed to establish a pressure or concentration gradient to get rid of the volatiles and semi volatile contaminants from soil (Baker and Bierschenk 1995). By utilising this method, large number of contaminants can be removed in a short duration.

(B) Floatation Technology: Due to the differences in the surface properties of the contaminants and the soil, oil is separated out from the soil by the help of a gas–liquid-solid system.

(C) Ultrasonication: The contaminants from the soil get desorped and thus form strong oxidants and hydroxyl radicals. These are then utilised to remove the pollutants (Lim et al. 2016).

(D) Electro-Kinetic Remediation: Electrodes are embedded deep in the contaminated soil, which provide direct electric current. The fluid medium moves towards the cathode as a result of the voltage potential gradient generated, dragging the pollutants along with it (Sri Ranjan et al. 2006).

(E) Thermal Desorption: It employs the temperature changes to elevate the contaminants' vapour pressure, which further become volatilized and subsequently desorbed from the soil.

(F) Biochar: It has a large specific surface area, is porous, is carbon-enriched, and is biodegradable. Due to its surface adsorption, partition, and sequestration properties, biochar was used as a solution to remediate organically polluted soil (Tang et al. 2013).

10.2.1.2 Chemical Remediation Approaches

Chemical oxidation has the capacity to preprocess or quickly deposit soil pollutants. Chemical oxidation remediation uses oxidants that can quickly and completely chemically destroy petroleum pollutants. Chemical oxidation involves chemically converting pollutants into non-hazardous forms, such as biodegradable products or less toxic compounds that are more stable, immobile, or inert (Lim et al. 2016). Plasma Oxidation based on pulsed corona discharge and dielectric barrier discharge is considered as a highly competitive technology for remediation of pollutants from soils. Plasma are macroscopically electrically neutral aggregates composed of numerous ions, electrons, atoms, etc. During the generation of plasma, a strong oxidizing environment is created for oxidative decomposition of contaminants. In addition, photocatalytic degradation is also an effective for decomposition of polycyclic aromatic hydrocarbon in the soil. This technique utilizes semiconductor metal oxide as catalyst to degrade organic pollutant into small molecules directly.

Advantages and Limitations of Physio-Chemical Methods

Remediation strategies have become a current research focus especially in terrestrial environments. The bioremediation methods have a number of key advantages over the traditional physio-chemical technologies, the key advantages being they are simple to implement, environment friendly, applicable over large areas of soil, cost-effective and leads to almost hundred percent destruction of soil contaminants. The main advantages and limitations of the physical methods are as follows:

(a) In the soil vapour extraction method, along with efficient and quick removal of contaminants it can also provide aeration to cause microorganism growth. The method's efficiency is however, affected by the properties of the soil and PHCs and the operation conditions.
(b) Since the flotation technology can even remove very minute or light weight particles with slow settling velocities, it can be considered a straightforward, low-cost operational, and highly effective method. The disadvantage is that it leads to a production of huge amount of wastewater.
(c) Ultrasonication is a highly efficient method due to its accurate heating and intense agitation but it comes at the high cost of excessive energy utilization for generation of the ultrasonic waves.
(d) Electro-kinetics remediation has an advantage in speed of execution and low operational cost. But it proves to be ineffective in low concentration of contaminants in the soil.

(e) Thermal Desorption can destroy contaminants under high heat conditions and with the advantage of releasing little or no gas into the atmosphere. However, this method is only limited to the contaminants that can be volatilised.
(f) The functioning and effectiveness of the photocatalytic degradation process are influenced by the soil's moisture content, humic material content, and light absorption properties

10.2.2 Emergence of Bioremediation as an Eco-Friendly Alternative.

Bioremediation is regarded as an efficient alternative technology that has little or no harmful effect on environment and it is easily application along with being cost-effective. This procedure uses living things or their products, either naturally or intentionally, to degrade, detoxify, mineralize, or convert the toxins in contaminated environments (Gouda et al. 2018). Thus, only those microorganisms that can sustain themselves and grow in the PHC contaminated soil are used for bioremediation. It is a feasible and sustainable technology. Bacteria, fungi and yeast are among the common organisms that are employed which important in the particular soil niche and play a vital role in bioremediation because of their high capacity to degrade contaminants in the soil. The most common species used are bacteria, which degrade petroleum compounds on a wide range of scales quickly. Similarly, plant-based approaches like phytoremediation, rhizoremediation, bio-stimulation, bio-augmentation, etc. are significant towards petroleum contamination management in environment. The phytoremediation makes the use of plants and microbes associated with the roots of these plants to degrade the contaminants. The selection of a plant for use as a phytoremediator should primarily take into account its root organisation and pattern, plant survival, and capacity to adapt to current environmental conditions. Plants can enzymatically degrade the contaminants or stimulate the rhizosphere and the microorganisms surrounding it to cause the degradation. In phytoremediation, the utilisation of plant growth-promoting rhizobacteria (PGPR) is necessary which gives phytohormones and mineral nutrients for plant growth. Additionally, PGPR can produce antibiotics, compete with pathogens for resources, or cause systemic resistance in the host plant to defend it against diseases (Chen and Zhong 2019).

10.2.3 Microbial Metabolism and Its Role in Hydrocarbon Degradation

According to numerous studies, one of the most effective ways to convert hydrocarbon depositions into less toxic and mobile forms or to mineralize these hydrocarbon contaminants into H_2O, CO_2, and microbial biomass is through microbial bioremediation, which uses microorganisms like bacteria, fungi, and microalgae (Das and Chandran n.d.).

Due to the introduction of hydrocarbon pollutants into the environment, microorganisms that can survive and multiply using these pollutants as their only source of carbon and energy have evolved. Even though these bacteria have the ability to breakdown hydrocarbon pollutants, there are a number of biotic and abiotic variables that limit how effectively toxins can be biodegraded.

10.2.3.1 Factors Influencing Microbial Oxidation

Utilization of a single strain or a consortium of indigenous bacteria has been shown to be effective for degradation in the laboratory scale. However, its use becomes limited in the field. This is due to many prevailing factors that can affect the efficiency and the rate of degradation by these microbes.

An overview of the major factors that affect bacterial degradation of petroleum contaminants in the soil are shown in Fig. 10.1. The major factors are classified as the biotic or abiotic factors as listed below. These factors alter microbial activities, the activities of enzymes having degradative action, and hydrocarbon degradation in general. This suggests that bacterial bioremediation might be more efficient and successful if these variables are adjusted, improved, and controlled.

10.2.3.2 Biotic Factors

The Availability of Microorganisms that Break Down Hydrocarbons

In PHC-contaminated soil, bacteria that break down hydrocarbons, either aerobically or anaerobically, are extremely common and all-pervasive. This is so that the bacteria can grow and readily adapt in the hydrocarbon-contaminated environments by using the pollutants as a source of energy and carbon. The rate at which hydrocarbon is degraded is directly linked with the presence of certain microorganisms that naturally found in the soil ecosystem (Macaulay 2015). The nature and the type of PHC depositions present is related to the species abundance and richness. Thus, the need for naturally occurring suitable bacteria that can implement its remedial actions.

Bacterial Competition and Cooperation

The hydrocarbon degrading bacterial species are present in the contaminated soil either in tandem with other species (synergistic effect) or in competition with them (antagonistic effect) (Chaerun et al. 2004). However, one limiting factor for the efficiency of the degradation process can be this inter-species or intra-species competition. Another factor can be that these microbial species when undergoing reactions with the PHCs can produce certain metabolites that will hinder the growth of other bacterial species (Galitskaya et al. 2016). The biodegradation rate of polyaromatic hydrocarbons is accelerated by the synergistic interaction between *Cycloclasticus*

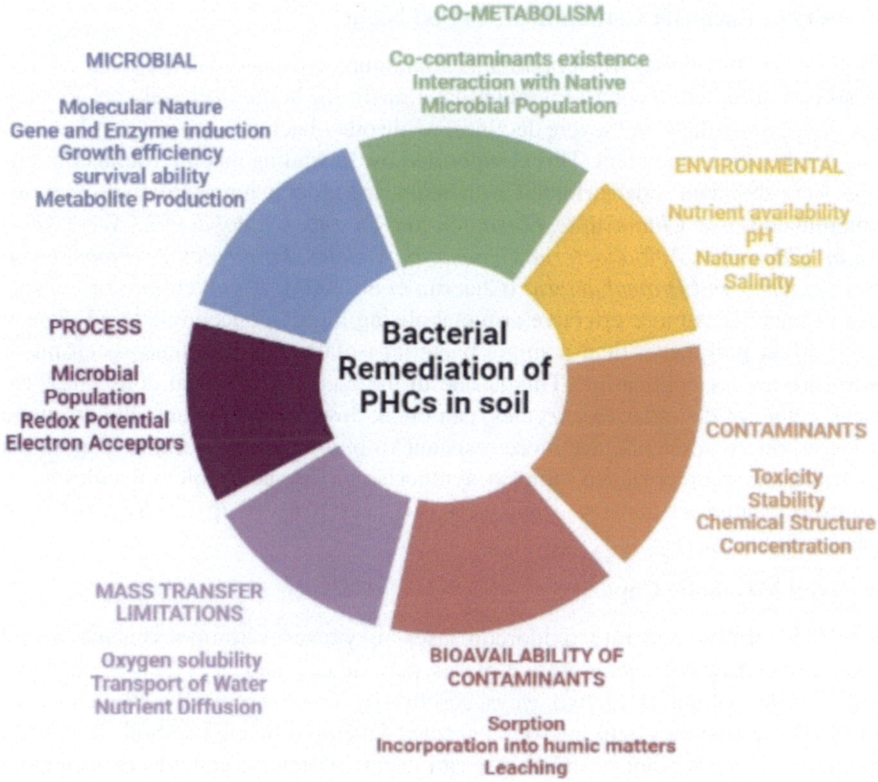

Fig. 10.1 Factors affecting bacterial remediation in petroleum contaminated soil

and *Alcanivorax borkumensis* and is slowed down by the antagonistic interaction between *Thalassolituus* and other oleophobic bacteria. The synergistic effect comes due to *A. borkumensis*, which uses other petroleum hydrocarbon contaminants as a metabolite, so it does not compete with PAHs for a carbon source, thus increasing the emulsification of PAHs (polycyclic aromatic hydrocarbons) makes them more easily accessible for *Cycloclasticus*. However, *Thalassolituus* releases comp0unds that can hinder the growth process of other oleophilic bacteria.

Exogenous and Indigenous Hydrocarbon-Degrading Bacteria

Microbes that are either present or newly introduced may help remove petroleum contaminants of soil. However, for the prolonged efficacy of the degradation process, native bacterial populations (indigenous and autochthonous), which may be isolated from hydrocarbon-contaminated settings, are more capable than foreign inocula (exogenous and allochthonous). Due to the exogenous consortia's inability to survive abiotic stress conditions, native bacteria demonstrated more effective biodegradation.

Diversity of Bacterial Communities in Soil Niche

Studies have shown that when pollutants containing hydrocarbons are released into the soil environment, bacterial populations transform. Some bacterial species may experience immediate and severe declines, while other bacterial species may become considerably more abundant. This is supported by the finding that microbial communities were abundant and distributed in an increasing order in long-term hydrocarbon-contaminated sites: *Chlamydiae, Elusimicrobia, Thermi, Cyanobacteria, Verrucomicrobia, Chloroflexi, Acidobacteria, Gemmatimonadetes, Firmicutes, Actinobacteria, Bacteroidetes, and Proteobacteria* (Chaerun et al. 2004). A polyculture or consortium of bacteria is more effective at metabolising highly concentrated and diverse hydrocarbon pollutants, than a single bacterial isolate with a limited spectrum of hydrocarbons for application. This is due to the fact that bacterial consortia have a wide range of degradative enzymes, can break down both aliphatic and aromatic hydrocarbon components, are more resistant to physical and chemical conditions (such as temperature, pH, and salinity), synthesise surface-active biomolecules (such as biosurfactants) at a faster rate, and work in concert to speed up the bioremediation process (Table 10.1).

Bacterial Metabolic Capacity

Microbial species generate oxidoreductases, oxygenases (monooxygenases and dioxygenases), dehydrogenases, hydrolases, peroxidases, and laccases to break down hydrocarbon pollutants in two ways aerobically or anaerobically (Beilen et al. 2004). These enzymes help microbial species develop efficient catabolic activities. In different types of gram-positive and gram-negative bacteria, genes have been identified that are essential for the bioremediation process. These genes include *assA*,

Table 10.1 Bacterial strains capable of utilizing long-chain n-alkanes as sole carbon source

Sl no	Microbial strains involved in degradation	Reported n-alkane chains	References
1	Bacillus thuringiensis A2	C_6–C_{28}	Throne-Holst et al. (2006)
2	*Alcanivorax borkumensis* AP1	C_{10}–C_{20}	Wang et al.(2006)
3	*Acinetobacter sp.* DSM17874	C_{10}–C_{40}	Hao et al. (2004)
4	*Geobacillus thermodenitrificans* NG80-2	C_{15}–C_{36}	Naik and Sakthivel (2006)
5	*Thermus sp.* C2	C_9–C_{39}	Yuste et al. (2000)
6	*Pseudomonas sp.* PUP6	C_{12}–C_{28}	Hamzah et al. (2017)
7	*Pseudomonas aeruginosa* RR	C_{12}–C_{34}	Imam et al. (2019)
8	*Xylella fastidiosa* RR15	C_{14}–C_{34}	
9	*Rhodococcus sp.* RR12 and RR14	C_{14}–C_{34}	
10	*Burkholderia cepacia* RR10	C_{12}–C_{34}	

bssA, phe, and PAH-RHDα among many more. Energy is required by bacteria for reproduction and biological processes. This energy is physiologically generated by a redox reaction within bacterial cells. As a result, bacteria use hydrocarbon contaminants as indicated in Table 10.2, as an energy source by catalysing the electrons transfer during aerobic or anaerobic bioremediation from their donors to acceptors.

10.2.3.3 Abiotic Factors

Hydrocarbon Characteristics

(a) Nature of Contaminants: A single bacterial strain or bacterial communities may be affected by the biodegradation, transportation, and metabolism of hydrocarbon pollutants depending on their occurrence, heaviness, viscosity and diffusion rate which are the physical characteristics and certain chemical characteristics like molecular weight, components and their structures. This is because of the hydrocarbons' molecular dimensions, make-up, structure, concentration, toxicity, and unexpected byproducts in the contaminated environment. These circumstances have an impact on the incidence, stability, and biological activity of bacteria that break down hydrocarbons.

(b) Concentration of Hydrocarbons: The rates of bacterial biodegradation's absorption, transformation, and mineralization are dependent on the pollutant quantities. High concentration pollutants are known to be dangerous and to adversely influence the rate of development and biomass generation of degraders due to their heft and undispersed nature. They also demand lengthy treatment times. Additionally, hydrocarbon concentrations greater than 5% can disturb the oxygen availability and C: N: P ratio in addition to reducing microbial breakdown activity. Similar to this, very low quantities of hydrocarbons prevent the production of breakdown enzymes by bacterial metabolic genes and limit the availability of carbon to sustain microbial development.

Physical Parameters

a. Characteristics of Soil: The soil provides the most varied habitat for different populations and types of microorganisms. The physical and chemical characteristics of the environment determine the makeup of the bacterial community, the type of native microorganisms, and the functional genes.
b. Soil Region and Particle Size: Due to the constant supply of organic matter (from plants and animals) and enough oxygen supply, the topsoil layer (vadose zone, soil surface, and unsaturated region of soil) typically harbours an enormous number of bacteria. In contrast, there is lesser quantity of bacteria in the saturated zone of the soil, i.e., the sediments because oxygen availability and organic matter supplementation diminish as soil depth increases. The permeability and rate of biological degradation of the soil are affected in tandem by the shape and size of the soil particle. Whereas coarse soil particles transmit these hydrocarbon contaminants through the soil to the unsaturated zone, tiny particles of soil (clay

Table 10.2 Molecular methods to monitor dynamics of microbial population and composition at the time of hydrocarbon bioremediation

Techniques	Principle	Dis-advantages
PCR: simple, real time and multiplex	Amplification and quantification is specific to target genes of interest	Contamination of DNA, non-specific primer annealing, and the existence of PCR inhibitors
Fluorescent in situ hybridization (FISH)	After treating microbial cells with fixatives, complete genomic DNA is hybridised with strain-, species-, or domain-specific DNA probes, improving the characterisation of uncultured microorganisms	Fewer probes might be used in a single hybridization experiment, and background fluorescence in some samples can interfere with epifluorescence or confocal laser imaging, which can impact the identification of microbes
Stable isotope probing (SIP)	Stable isotope-labelled substrates may be incorporated into cellular biomarkers including rRNA, PLFA, and DNA to help identify the organisms that are consuming the substrates. It ties phylogeny and ecological roles together	Insensitive; enrichment bias could not accurately represent substrate metabolism in the environment
Automated ribosomal intergenic spacer analysis (ARISA)	Useful for metagenomic studies Polymorphism in the length of fluorescently labelled ITS regions	For complete organism identification, the database is less comprehensive than the 16S database
Temperature gradient gel electrophoresis/Denaturing gradient (TGGE/ DGGE)	On the basis of the separation of partly melted 16S rRNA in a linear temperature/denaturing gradient, polymorphism Bands can be removed from a gel as well as sequenced for phylogenetic study and identification	Only 500 bp of the 16S rRNA sequence from the microbial community is available for sequence analysis
Terminal restriction fragment length polymorphism (T-RFLP)	Modification of ARDRA; however, the fluorescently labelled PCR primers used in T-RFLP now allow for the visualisation and measurement of the PCR products The 16S gene's polymorphism in the size of the fluorescently tagged terminal restriction fragment aids in the enumeration of the microbial population	For the best description of the microbial community in a sample, many restriction enzymes are required

(continued)

Table 10.2 (continued)

Techniques	Principle	Dis-advantages
Amplified ribosomal DNA restriction analysis (ARDRA)	Restrictions enzymes are used to break down PCR-amplified 16S rRNA segments to produce a microbial community profile	The selection of restriction enzymes has a significant impact on the microbial community fingerprint
Single strand conformation polymorphism (SSCP)	Polymorphism that is based on single-stranded 16S rRNA in polyacrylamide gel	Covers not more than 500 bp of 16S gene
16S pyrotags	High-throughput metagenomic analysis, Pyrosequencing of 16S rRNA	Can only provide a 200 kb short tag for the 16S gene
Reverse sample genome probing (RSGP)	Genomes that show over 70 percent cross hybridization are frequently thought to belong to the same species. cross-hybridization between pure-culture extracted chromosomal DNA of microbes and typical microbe species. used to classify and identify microorganisms	Not appropriate for environments where pure culture separation has not previously taken place; not relevant for environments with uncultured organisms

soil; with smaller interstitial areas) retain the pollutants at the soil surface thereby decreasing the availability of nutrients and oxygen. As a result, degradation rate decreases extensively. Abstemiously washed soil particles are essential for raising the bioavailability of the contaminant, supplying oxygen, and stimulating the metabolism and growth of local microbes (Geng et al. 2020).

c. Aeration/Oxygen Availability: The limitation of aerobic bioremediation is molecular oxygen in the soil, which also functions as the ultimate electron acceptor (Xu et al. 2018) and a chemical reactant (which consists of enzymatic oxygenase reactions) in aromatic hydrocarbon contaminants' terminal, subterminal, as well as bi-terminal oxidation and ring cleavage. According to a study, 3.1 mg/ml of oxygen is necessary to break down 1 mg/ml of hydrocarbon pollutants, and 10–40% of oxygen is needed for successful hydrocarbon biodegradation. This is without accounting for the complete biomass of prospective bacteria that can degrade hydrocarbons (Sihag et al. 2014). Thus, aerobic catabolism has a higher rate of hydrocarbon degradation as compared to metabolism carried out anaerobically.
d. Effect of Temperature: Temperature has a considerable impact on the enzyme activity, rate of bacterial growth, the physiology, the chemistry of the pollutants, and presence of bacterial species diversity responsible for the degradation processes. Temperature variation affects the bacterial population that cause the hydrocarbon degradation, and moderate temperatures are typically responsible for the selection of mesophilic bacteria that degrade contaminants. The broad array of performance of several hydrocarbon-degrading bacteria was also

demonstrated, highlighting the significance of temperature in the remediation of hydrocarbon contamination (Lu et al. 2011).

a. Effect of pH: Most microorganisms that break down hydrocarbons prefer neutral to slightly alkaline pH levels. The development and accumulation of bacterial waste products in hydrocarbon-contaminated soils can change the pH, which has a bearing on the solubility of pollutants, the accessibility of nutrients, the activity of microbes, and bioavailability. As highlighted by some studies, minimal degradation of hydrocarbons take place in a low acidic pH range of 2–5.5 and alkaline pH 7.5–10 (Wu et al. 2017), while efficient degradation is obtained at nearly neutral pH (5.2–7.0) whereas the maximum population of bacteria is observed at pH 7.5.
b. Effect of Moisture: When soil is contaminated with hydrocarbons, it loses water-holding ability and porosity. Since bacteria and soil water activities are strongly correlated, this state ultimately reduces microbial activity. Decrease in the moisture content, bacterial population reduces, and conversely, oxygen transmission in inhibited when the moisture level in the soil is high. Due to the need for water for microbial development and metabolism, hydrocarbon breakdown in terrestrial ecosystems is constrained. Because of this, the right amount of moisture availability—between 50 and 75%, which is crucial for hydrocarbon biodegradation.
c. Effect of Salinity: When salt is deposited and accumulates in an ecosystem, whether naturally or as a result of human activity, halophile or halotolerant species are subjected to greater stress.. High salinity can cause an increase in osmotic pressure, impede microbial metabolic processes, impair their solubility and ability to transfer vital ions, and deprive them of oxygen. Halophilic bacteria have the capacity to endure in such hypersaline settings and breakdown the contaminants by creating hydrocarbonoclastic enzymes (Sierra-Garcia and Oliveira 2013).

10.3 Metabolic Pathways Involved in Microbial Oxidation

The substances that just contain carbon and hydrogen are known as hydrocarbons. The majority of non-polar hydrocarbons and are lesser prone to chemical reactions at ambient conditions of temperature because they lack functional groups. According to their bonding characteristics, hydrocarbons are typically categorised as follows:

(i) The aliphatic group, which consists of cyclic, branched-chain, and straight-chain (n-alkane) compounds.
(ii) the aromatic group, which comprises many significant compounds also consisting of aliphatic hydrocarbon chains (alkylbenzenes for example) and mono- or polycyclic hydrocarbons.

It has been thoroughly studied and determined that bacterial isolates can completely breakdown the hydrocarbons majorly, including aliphatic and aromatic compounds. This type of degradation can take place either anaerobically, without oxygen, or aerobically, with oxygen and other acceptors of electrons such ferric ion, nitrate or sulphate. Aliphatic vs. aromatic hydrocarbons, aerobic vs. anaerobic degradation pathways, physiology and general metabolic pathways, vs. enzymatic mechanisms and structures—these are the traditional divisions used to research biodegradation. the "metabolic challenge" of introducing an apolar, inert substance made exclusively of carbon and hydrogen into metabolism (Rojo 2010).

10.4 Microbiological Hydrocarbon Breakdown Processes Based on Biochemistry and Genetics

The enzymatic processes that are used in the aerobic hydrocarbon breakdown by bacteria have been the subject of extensive research for many years. For aerobic and easily cultivable bacteria, such as the Pseudomonas putida GPo1 strain, *Mycobacterium tuberculosis* H37Rv and Acinetobacter sp. ADP1, genes encoding degradation enzymes are rather well understood. The anaerobic hydrocarbon degradation, on the other hand, has drawn considerable attention because it is thought to be the main process at work in many polluted settings and oil reservoirs.

10.4.1 Aliphatic and Aromatic Petroleum Hydrocarbons Undergoing Aerobic Degradation

10.4.1.1 Aliphatic Hydrocarbons

In most of the degradation pathways described, the following takes place:

- In the activation step, the substrate is n-alkane.
- With the help of terminal and sub-terminal oxidation, it is converted by Substrate specific terminal monooxygenases/ hydroxylases to the necessary primary and secondary alcohols. Some bacteria may co-exist with both terminal and sub-terminal oxidation.
- Following the oxidation of the alcohol to the appropriate aldehyde, the fatty acid is subsequently transformed into a fatty acid and conjugated to acyl CoA.
- The -oxidation pathways break down acyl-CoA in the final phase of metabolism to produce acetyl-CoA (through the TCA Cycle).

The primary alcohol leads to aldehyde formation whereas the secondary alcohol leads to Ketone formation but ultimately, they form the Acyl-CoA to go through the β—oxidation pathway (Kuhn et al. 2009). For the initial terminal hydroxylation of n-alkanes in bacteria, various families of enzymes, including:

(1) propane monooxygenase (C_3),
(2) CYP153 monooxygenases (C_5-C_{12})
(3) *AlkB*-related non-heme iron monooxygenase (C_3-C_{10} or C_{10}-C_{20}),
(4) different classes of butane monooxygenase (C_2-C_9)
(5) flavin-dependent monooxygenase *LadA* (C_{10}-C_{30}),
(6) copper flavin-dependent dioxygenase (C_{10}-C_{30}),
(7) flavin-binding monooxygenase *AlmA* (C_{20}-C_{36}),

One of the most well researched and best alkanes activating enzyme is integral membrane non-heme iron monooxygenase (*AlkB*). In several environmental samples and bioremediation experiments, these genes have been widely employed as functional indicators or rather biomarkers for the identification of aerobic alkane-degrading bacterial communities because they have been frequently linked to the presence of microorganisms that break down long (>C_{12})- and medium (C_5–C_{11}) length alkanes (Salminen et al. 2008) (Fig. 10.2).

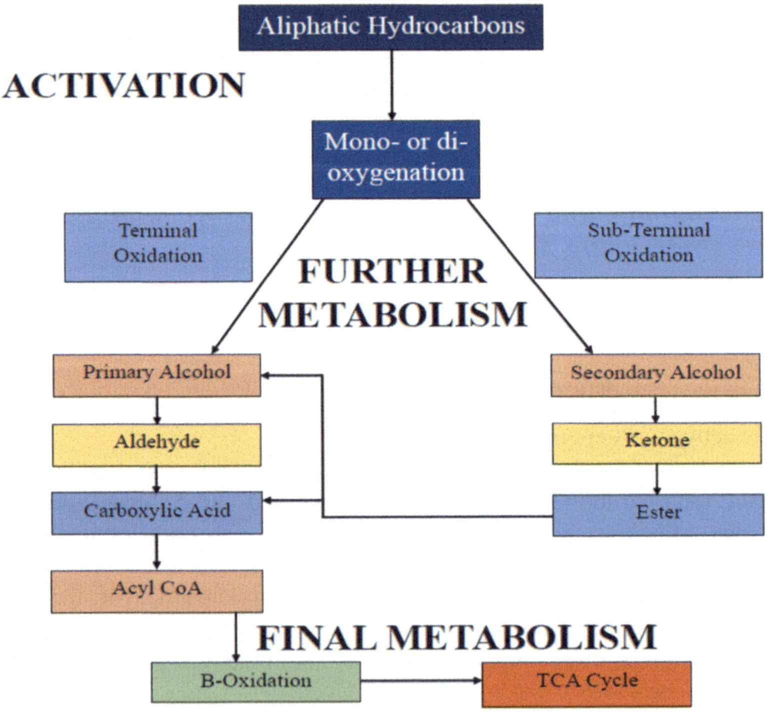

Fig. 10.2 The Aerobic pathway for the degradation of aliphatic hydrocarbon

10.4.1.2 Aromatic Hydrocarbons

These aromatic compounds are catabolized by the aerobic bacteria thus involving a large number of peripheral routes that convert structurally diverse substrates into a small number of common intermediates, that are next processed and degraded by some central pathways. Characterization of the metabolic pathways and genes involved for the breakdown of certain members of a very wide spectrum of aromatic compounds has been done using isolated bacterial strains, especially from *Proteobacteria* as well as *Actinobacteria* phyla (Guzik et al. 2013).

Degradation by such isolates is often begun by members of one of the three superfamilies:

(1) the soluble diiron multicomponent monooxygenases (SDM).
(2) the Rieske non-heme iron oxygenase (RNHO),
(3) the flavoprotein monooxygenases (FPM) and

The aerobic degradation pathway for the aromatic hydrocarbons is as follows:

- CoA ligases carry out the activation step where the generated CoA derivates undergo selective hydroxylation.
- Then, mono- or dihydroxylation of the aromatic rings take place.
- Due to peripheral hydroxylation reactions, these are further metabolised and the rings get cleaved either by Intradiol or Extradiol dioxygenases (INDO) or (EXDO).
- These then follow some central pathways reactions and finally enter the β—oxidation pathway, leading to the TCA cycle.
- Another method is that the aromatic hydrocarbons enter the Co-A thioester pathway where the Co-A thioesters get reduced and these also ultimately end in the β—oxidation pathway.

As a way to start the aerobic degradation of hydrophobic pollutants such as benzene, toluene, naphthalene, biphenyl, or polycyclic aromatics, the aromatic ring is usually activated through oxygenation reactions supported by RNHO enzymes or, as strongly described for toluene degradation, by means of members of SDM enzymes (Widdel and Grundmann 2010).

10.4.2 Anaerobic Degradation of Aliphatic and Aromatic Hydrocarbons

10.4.2.1 Aliphatic and Aromatic Hydrocarbons

Since the BTEX (benzene, toluene, ethylbenzene and xylene) compounds are classified as pollutants of first priority, the aliphatic hydrocarbons are not studied as widely. Additionally, in case of the n-alkanes their anaerobic growth and solubility is less. However, n-alkanes can also be equally toxic and difficult to remove, hence their degradation is also necessary (Carmona et al. 2009). It has been shown that many anaerobic bacteria may degrade n-alkanes with six or more carbons, notably hexadecane (C16), by using sulphate or nitrate to act as electron acceptor (Kube et al. 2004). Two main methods of anaerobic degradation have been characterised.

The first method:

- Fumarate is added to the sub-terminal carbon of the alkane by the help of glycyl-radical enzymes (like 1-methyl-alkyl succinate synthase or alkyl succinate synthase).
- This leads to the formation of alkyl succinates. There have been reports of alkyl succinates derived from C_3 to C_{11} alkanes and suspected metabolites derived from substances with one degree of unsaturation, which include alkenes or alicyclic alkanes.
- Carbon skeleton rearrangement of the alkyl succinates take place which leads to their de-carboxylation and finally these go into the β—oxidation pathway.

The second method:

- Employs a process of carboxylation, derived from the growth pattern of Hxd3 (*Desulfococcus oleovorans*), a sulphate-reducing strain.
- The compounds underwent further metabolization to form C-even and C-odd n-alkanes.
- The C-odd n-alkanes were transformed into C-even cellular fatty acids by this strain Hxd3, while the C-even n-alkanes were transformed into C-odd cellular fatty acids. The Hxd3 strain is different from the other alkane degrading strains in this way.
- Finally, the fatty acids formed go into the β—oxidation pathway.

Microorganisms attack the aromatic ring using a totally different pathway in the absence of oxygen, one based on reductive processes. Recently, six bacterial strains which can act on aromatic hydrocarbons, anaerobically and break them down using numerous acceptors of electrons and belonging to different taxonomic strains (Rodrigues et al. 2009), have been identified and sequenced.

a. *Azoarcus* sp. EbN1 and *Thauera aromatica*,—Denitrifying betaproteobacteria
b. *Magnetospirillum magneticum* strain AMB-1, the denitrifying strain and *Rhodopseudomonas palustris* strain CGA009, a phototroph bacterium—Two alphaproteobacteria
c. *Syntrophus aciditrophicus* strain SB. the fermentor strain,
d. *Geobacillus metallireducens* GS-15, the iron reducer- two obligate anaerobic deltaproteobacteria.

10.5 Monitoring and Assessing the Effectiveness of Microbial Degradation.

Monitoring and assessing the effectiveness of microbial degradation is a critical endeavour in the realm of environmental science and biotechnology. Microbial degradation plays a pivotal role in remediation efforts, wastewater treatment, and the bioremediation of polluted environments. The ability to gauge and optimize the efficiency of these microbial communities is paramount for achieving desired outcomes, be it the removal of pollutants, the degradation of organic waste, or the sustainable production of biofuels. This multifaceted discipline encompasses a range of techniques and analytical tools that enable scientists and engineers to measure, track, and ultimately enhance the performance of microbial populations in various applications. In this context, the assessment of microbial degradation effectiveness stands at the forefront of environmental stewardship and the pursuit of greener, more sustainable solutions to pressing global challenges.

10.6 Techniques for Analysing Microbial Diversity in Soil Undergoing Bioremediation

For bioremediation to be regarded as a successful and effective procedure, it is critical for an adequate no. of active microorganisms to be present in the soil environment, that is capable of attacking the hydrocarbon pollutants. If the site has adequate amount of water content, favourable pH range, and has only one type of PHC contamination, then the possibility of the presence of an active population of bacteria degrading hydrocarbons, is extremely high. This means that bioremediation will also succeed (Philp et al. 2005). The initial steps in monitoring involve crucial chemical tests required for recognizing the pollutant and the commonly used microbiological methods for quantifying viable microbe populations. The methods used for quantifying microbes that consume hydrocarbons are common, affordable, and generally quick. However, even when it challenging for the development of the population of a specific microbe, the method of developing a mixed culture that contain significant PHC degrading bacteria, has shown to be more beneficial. The viable but nonculturable issue has led to the development of microbial ecology approaches that are not dependant on culturing and have been shown to be extremely helpful for tracking the progress of bioremediation.

10.6.1 *Microbiological Methods*

10.6.1.1 Hydrocarbon-Utilizing Bacteria (HUB) Monitoring

Crude oil and other PHCs are heterogenous chemically and are present not only in sites of oil pollution, but also present in the most pristine soils and sediments, as a result of biosynthesis by plants and microorganisms and seepage from natural deposits. HUB are extensively present in the nature and there abundance is seen in environments polluted with PHCs which are presently under bioremediation process (Wrenn and Venosa 1996). However, the diversity of the different bacterial strains present can decrease to a great extent even when there in an increase in the percentage of HUB. This happens because the hydrocarbons present naturally often lead to enrichment of HUB in a selective manner (Chikere et al. 2011).

The nutritional requirements which act as rate-limiting factors, must be met for bacteria to utilise hydrocarbons. These growth factors include metals like K^+ and Na^+ as well as nitrogen, phosphorus, and sulphur for the oxygenases. The hydrocarbons under consideration will determine the technique used for HUB isolation.

Enumeration and Isolation of Culturable HUB

As a result, the requirement for an mineral salt agar-based media for separation and isolation serves as the foundation. One highly helpful mineral salt media for HUB enumeration and isolation is the Bushnell Haas (BH) agar, which a few researchers have employed it for this purpose. The vapour phase transfer method is frequently used to isolate bacteria utilising crude oil, taken from PHC contaminated soil. In this method, an inoculation of a known soil suspension sample is used and added to the BH agar, and a crude oil saturated filter paper is placed in the Petri dish lid. One evident drawback of this method, according to Philp et al. (Itävaara and Vikman 1996), is that it can only be used to isolate a subset of the overall population of bacteria that can use the hydrocarbons of the vapour phase. The entire population of organisms that degrade crude oil may be underestimated by this method. It is advised that findings from the culture-dependent isolation of bacteria that use hydrocarbons be verified using culture-independent methods such as denaturing gradient gel electrophoresis (DGGE), using 16S rRNA genes that have been amplified using PCR.

Enumeration of Unculturable HUB

Only a percentage of the environmental samples can be cultured. Organisms that grow rapidly and can thrive in specific conditions dominate cultures, leading to an inaccurate representation of the actual microbial community. Methods dependant on cultivation may restrict microbial diversity assessment, potentially overlooking or underestimating the ecological significance of unculturable organisms. These primarily uncultured organisms' genomes code for a new enzyme with unexplored metabolic capabilities. Microbial cultivation is not a requirement for Metagenomics analysis. For the functional genomic characterization of hydrocarbon degraders, nucleic acid-based techniques are used, including polymerase chain reaction (PCR), denaturing gradient/temperature gradient gel electrophoresis (DGGE/TGGE), and other microbial typing techniques based on rRNA sequences. A table of the different molecular techniques used have been shown in Table 10.2.

MPN Techniques on Microtiter Plates

Due to the disadvantages of traditional isolation methods based on solid-agar; liquid based culture methods like the most-probable number (MPN) procedure were developed. In order to calculate the MPN of microbes in the initial samples, multiple replicates of the samples are analysed. This is a statistical method based on sample dilution to extinction. Wrenn and Venosa (Elsas 2006). To count aromatic and aliphatic hydrocarbon degraders individually in different plates, a 96-well microtiter plate MPN method was created. This strategy was created especially with crude oil as a substrate in mind. The n-hexadecane utilised in the alkane degrader MPN method is used as the carbon source, and turbidity is used to measure growth and iodonitrotetrazolium violet reduction to iodonitrotetrazolium fromazan (red precipitate) to measure electron transport activity. The absence of the hydrocarbon is an intrinsic shortcoming of the microtiter plate based MPN technique. The oil and aqueous phases are both in less quantity. This shows that the incubation periods are long, upto two weeks and there is a considerable chance of media contamination or drying out of the wells. Water salinity can increase due to high rates of evaporation (Hurst et al. 2007).

Use of Enzyme Activities to Isolate HUB

Due to their crucial functions in both aromatic ring cleavage and substrate activation, dioxygenases are the enzymes that are most relevant to the breakdown of hydrocarbons. Indole can be used to test for dioxygenase activity in agar plates with mineral salt, where Indigo is formed from Indole conversion by dioxygenases which give blue colonies that can be selected (Drenovsky et al. 2004).Catechol 2,3-dioxygenase screening is a more focused enzyme test.A very significant and frequent intermediary in the breakdown of aromatic hydrocarbons is catechol. Colonies on plates that have been treated with catechol can be sampled using a filter lift. A yellow pigment appears, after 10 min incubation times, thus indicating catechol 2,3-dioxygenase activity (Zelles 1999).

10.6.1.2 Physiochemical Techniques

Respirometry

The rate of the reaction is measured using metabolic gas respirometry by detecting the production of CO_2 or the Consumption of O_2. Radio respirometry utilising ^{14}C-labeled hydrocarbons which get labelled at the most obstinate part, is a more sensitive and accurate method. However, it is an expensive technique and requires a technical expertise to perform. Since oxygen consumption might result via biotransformation rather than always mineralization, it is less responsive than CO_2 evolution. Evaluation of biodegradability in solid mediums like soil and sediments using CO_2 generation provides invaluable information on mineralization (Maletić et al. 2019). Metabolic gas respirometry offers greater flexibility due to fewer technical limitations compared to other methods. Nevertheless, concerns regarding its sensitivity have arisen, particularly in locations with low oxygen concentrations. While an effective method for biokinetic parameters' detection in groundwater for degradation of pollutants, can be respirometry, its successful application in soil remains unproven.

10.6.1.3 Biochemical Methods

Community Level Physiological Profiling (CLPP)

An ecosystem's metabolic diversity/fingerprint (pattern of substrate utilisation) is expressed through community level physiological profiles. This is based on the substrates identified and that can be oxidised with the help of indigenous autochthonous bacteria. The diversity in this context is simply the capacity of the entire microbial community to breakdown certain substrates (range of carboxylic acids, sugars, peptides and aminos acids) to provide an abundant amount of precise biochemical assays.

By using the tetrazolium-based redox dye technique of Biolog (Biolog, Inc., Hayward, CA, USA), the CLPP of utilization of carbon source can be examined. Gram-positive and Gram-negative bacteria are grown in microplates (having 96 wells) or ecoplates that have distinct carbon sources in the range of 31–95. Thus, a measurement of the community metabolic capacity can be obtained from the distinctive substrate utilisation model, or "metabolic fingerprint," formed by the community of microbes after being isolated from any environmental sample like soil sample (Maletić et al. 2019). The CLPP approach has drawbacks because it depends on the culture and does not give any idea about the structural information which is more effective for microbial diversity studies. Instead, it provides functional information on microbes present in the contaminated environments.

Phospholipids Fatty Acids (PLFA) Analysis

To examine the architecture and metabolic states of microbial communities, as well as to identify and contrast the differences and similarities among the microbial communities, total phospholipids fatty acids were isolated from environment like soil samples (Maletić et al. 2019). Odd-numbered and branched-chain fatty acids are linked to Gram-positive bacteria, whereas eubacteria typically do not contain polyunsaturated fatty acids and anaerobic prokaryotic bacteria have plasmalogen enriched phospholipids. Gram-negative bacteria possess greater amounts of cyclopropane fatty acids as well as even-numbered monosaturated straight chain (Koshlaf and Ball 2017).

In reaction to environmental factors including pollution, chemical stress, and temperature changes, microorganisms can alter the lipid content of their membranes. Cell death causes rapid degradation of PLFA, thus makes it a reliable indicator of living organisms (Galitskaya et al. 2016). Additionally, variations in PLFA patterns in response to environmental stress can be used as biomarkers to characterize the physiological state of certain microbial taxa and its community structure.

When interpreting changes in phospholipid profiles, one can make use of an online database of pure cultures and established biochemical pathways. Variations in phospholipid profiles are typically associated to variations in the proportion of microbial groups (Kumar et al. 2021). Gas chromatography equipped with a mass spectrometer is used to quantitatively examine the extracted fatty acids, and data comparison with a database of fatty acids enables the identification of the isolated PLFAs. Although species-level delineation cannot be achieved with the direct removal of PLFA from soil, this method is effective for profiling broad alterations in microbial community structure (Kumari et al. 2022).

10.6.2 Optimization of Bioremediation Strategies

It is necessary to optimise the biotechnological process in order to assess the numerous factors that affect the process' performance and cost. One of the crucial elements to optimise for bioremediation of pollutants is the amount of microbial biomass required, as its production will directly affect the process' cost. Herbicides and non-chlorinated pesticides, heavy metals (chromium, lead, radionuclide, mercury etc.), nitrogen compounds and organic and inorganic pollutants (hydrocarbons, halogenated organic solvents and compounds, herbicides and non-chlorinated pesticides, etc.) can all be removed or reduced using bioremediation methods with low operational costs. Different macro- and microorganisms are employed for this application, but bacteria are particularly helpful because they are widespread and used for ex situ or in situ bioremediation processes. They selectively breakdown a range of hydrocarbon contaminants. The various bioremediation techniques can be broadly divided into three categories: natural attenuation, bioaugmentation, and bio stimulation.

Biostimulation: *Stimulating microbial activity through nutrient amendments.*

In hydrocarbon-contaminated environments, biostimulation entails the inclusion of growth-restraining elements including substrates, vitamins, and oxygen as well as the modification of the natural circumstances (temperature, pH, moisture, terminal electron acceptors, redox potential, etc.) (Maletić et al. 2019). This is crucial to enhancing the metabolic capacity of native bacteria that break down hydrocarbons. When hydrocarbon pollutants are widely dispersed, they offer significant levels of carbon and energy but little in the way of phosphorus or nitrogen for bacterial growth. Thus, the introduction of these components increases the nutritional level needed to promote the growth of local bacteria. However, there are several restrictions on the use of bio stimulation. These include the soil's makeup, the lack of naturally occurring microbes that break down hydrocarbons, and the rapid depletion of important inorganic nutrients (NP). When native or naturally occurring, microbial species cannot breakdown pollutants quickly enough, microbial activity can be boosted by adding fertilisers

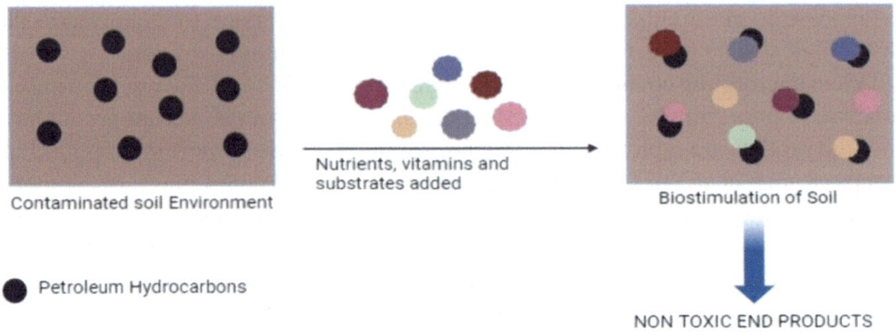

Fig. 10.3 Bio stimulation of microbial activity for enhanced bioremediation

which have nutrients in them, such as $NaNO_3$, $MgNH_4PO_4$, KNO_3, K_2HPO_4, and NH_3NO_3.(Atashgahi et al. 2018). By taking into account the key elements determining the entire process of degradation in the environment, the concurrent utilization of bioaugmentation and bio stimulation generally results in significant and effective hydrocarbon bioremediation procedures (Fig. 10.3).

10.7 Current Status, Challenges and Future Prospects

Many different contaminants are removed using bioremediation. After being tested in the lab, the procedure is expanded to a larger area, i.e., a contaminated soil site. Each time, various adjustments and changes are to be made to the fundamental technologies for the process optimization. The factors are mainly dependant on the type, place and source of the contaminant. Therefore, it is vital to look at numerous insights of this phenomena in order to grasp the present scenario in light of the recent technologies and advancements in the bioremediation process. The key limitations that should be taken care of while soil bioremediation takes place are as follows:

- Pollutant bio-accessibility or bioavailability: Adsorption, toxicity, and utilisation of a pollutant are what determine its accessibility to soil bacteria. The soil contains numerous organic substances, like different pesticides and polycyclic aromatic hydrocarbons (PAHs), which are strongly toxic. Compared to a low concentration of PAH-degrading microbes, reduced bioavailability has a stronger effect on bioremediation of pesticide and PAH-polluted soil. Contaminants' bioavailability varies over time. This process is called ageing. One of the biggest difficulties in cleaning up soil contamination is the ageing of contaminants. Finding a suitable strategy to increase the bioavailability of contaminants in soil is therefore necessary.

- Pollutant stability reversion: One viable and sustainable strategy to address the buildup of organic products is to compost organic matter recovered after waste treatment. However, the environment's pH is altered when organic matter mineralizes, which has the opposite effect on pollutant stability.
- Microbial adaptability: Exogenous microorganisms' sluggish adaptability to work on a specific site is caused by their poor compliance with the polluted soils. As a result, bioaugmentation loses its ability to accelerate the breakdown of contaminants. It may be necessary to use appropriate approaches to investigate and solve this issue.
- Metabolic pathways: Currently, little is known about the functioning of microbial metabolism during the bioremediation of organic pollutants and heavy metals. In order to genetically modify foreign microorganisms, in-depth study of the metabolic processes and the rhythms of microbial populations is therefore crucial. To understand the genetic organisation of indigenous bacteria, molecular biology tools like high-throughput DNA sequencing should be applied. To get beyond the aforementioned restrictions and remove the toxins from the environment, synthetic biology methods and technology are required (Yesilada et al. 2018).
- Enzymatic studies: Basic catalytic activity of various enzymes hides the degradation of pollutants, significant research is needed to understand a number of enzymatic characteristics, such as molecular structure, kinetics, inhibition mechanism, and activity of the enzyme (Tortella et al. 2015).
- Interactions with pollutants: The mechanism of how pollutants coming from the compost/ soil mixtures interact with one another to affect how quickly they degrade, is not studied yet. Further research is necessary to overcome this obstacle.
- Quality of end product: The final product that is released during bioremediation must be free of all forms of dangerous chemical substances and minerals that are present over a threshold level.

10.8 Conclusion and Future Prospects

For the treatment of contaminated aquifers and soil, bioremediation has become an increasingly prevalent method. Currently, effectiveness related to specific microorganisms in degrading environmental pollutants is widely acknowledged. However, challenges arise when attempting to transition this microbial capability from laboratory settings to real-world field applications. Nonetheless, the scientific community continues to make relentless efforts in this domain, pushing the boundaries of existing bioremediation techniques. Researchers worldwide are actively seeking innovative engineering solutions to enhance the microbial activity. For instance, recent developments such as gas sparing have significantly improved the aerobic degradation of petroleum products. The researchers are devising new methods to enhance microbial bioavailability, thereby increasing the bioremediation efficiency. Techniques such as waste solubilization through heat injection using steam, hot air, or flushing with hot water, subsurface matrix fracture with intense-pressure, and the introduction of surfactants are among the approaches for improving the current bioremediation strategies. As the field of bioremediation progresses, it is essential not only to discover effective and feasible strategies but also to develop means of evaluating the efficiency and functionality of current methods.

Acknowledgements The authors acknowledge their department and Institute for providing essential facilities and support.

Author Contributions SD, SB and SS contributed equally to the conception of the study, manuscript writing, and editing. NP contributed significantly to the manuscript writing, and illustration preparation. RP and SKP wrote the manuscript with constructive discussions.

Consent for Publication All authors have consented to the manuscript for publication.

References

Atashgahi S, Sánchez-Andrea I, Heipieper HJ, van der Meer JR, Stams AJ, Smidt H (2018) Prospects for harnessing biocide resistance for bioremediation and detoxification. Science 360(6390):743–746

Baker RS, Bierschenk J (1995) Vacuum-enhanced recovery of water and NAPL: concept and field test. J Soil Contamination 4(1):57–76

Beilen JB Van, Marín MM, Smits TH, Röthlisberger M, Franchini AG, Witholt B, Rojo F (2004) Characterization of two alkane hydroxylase genes from the marine hydrocarbonoclastic bacterium *Alcanivorax borkumensis*. Environ Microbiol 6(3):264–273

Carmona M, Zamarro MT, Blázquez B, Durante-Rodríguez G, Juárez JF, Valderrama JA, Barragán MJ, García JL, Díaz E (2009) Anaerobic catabolism of aromatic compounds: a genetic and genomic view. Microbiol Mol Biol Rev 73(1):71–133

Chaerun SK, Tazaki K, Asada R, Kogure K (2004) Bioremediation of coastal areas 5 years after the Nakhodka oil spill in the Sea of Japan: isolation and characterization of hydrocarbon-degrading bacteria. Environ Int 30(7):911–922

Chen S, Zhong M (2019) Bioremediation of petroleum-contaminated soil. Environ Chem Recent Pollut Control Approach 27(34):1–2

Chikere CB, Okpokwasili GC, Chikere BO (2011) Monitoring of microbial hydrocarbon remediation in the soil. 3 Biotech 1:117–38

Drenovsky RE, Elliott GN, Graham KJ, Scow KM (2004) Comparison of phospholipid fatty acid (PLFA) and total soil fatty acid methyl esters (TSFAME) for characterizing soil microbial communities. Soil Biol Biochem 36(11):1793–1800

Galitskaya P, Akhmetzyanova L, Selivanovskaya S (2016) Biochar-carrying hydrocarbon decomposers promote degradation during the early stage of bioremediation. Biogeosciences 13(20):5739–5752

Geng S, Cao W, Yuan J, Wang Y, Guo Y, Ding A, Zhu Y, Dou J (2020) Microbial diversity and co-occurrence patterns in deep soils contaminated by polycyclic aromatic hydrocarbons (PAHs). Ecotoxicol Environ Saf 15(203):110931

Gouda S, Kerry RG, Das G, Paramithiotis S, Shin HS, Patra JK (2018) Revitalization of plant growth promoting rhizobacteria for sustainable development in agriculture. Microbiol Res 1(206):131–140

Guzik U, Hupert-Kocurek K, Wojcieszyńska D (2013) Intradiol dioxygenases—the key enzymes in xenobiotics degradation. Biodegradation Hazardous Spec Products 14(7):129–153

Hamzah A, Manikan V, Abd Aziz NA (2017) Biodegradation of tapis crude oil using consortium of bacteria and fungi: Optimization of crude oil concentration and duration of incubation by response surface methodology. Sains Malaysiana. 46(1):43–50

Hao R, Lu A, Wang G (2004) Crude-oil-degrading thermophilic bacterium isolated from an oil field. Can J Microbiol 50(3):175–182

Hurst CJ, Crawford RL, Garland JL, Lipson DA (eds) (2007) Manual of environmental microbiology. American Society for Microbiology Press

Imam A, Suman SK, Ghosh D, Kanaujia PK (2019) Analytical approaches used in monitoring the bioremediation of hydrocarbons in petroleum-contaminated soil and sludge. TrAC, Trends Anal Chem 118:50–64

Itävaara M, Vikman M (1996) An overview of methods for biodegradability testing of biopolymers and packaging materials. J Environ Polym Degrad 4:29–36

Koshlaf E, Ball AS (2017) Soil bioremediation approaches for petroleum hydrocarbon polluted environments. AIMS Microbiology. 3(1):25

Kube M, Heider J, Amann J, Hufnagel P, Kühner S, Beck A, Reinhardt R, Rabus R (2004) Genes involved in the anaerobic degradation of toluene in a denitrifying bacterium, strain EbN1. Arch Microbiol 181:182–194

Kuhn E, Bellicanta GS, Pellizari VH (2009) New *alk* genes detected in Antarctic marine sediments. Environ Microbiol 11(3):669–673

Kumar V, Kaushal A, Shah MP, Singh K (2021) Phytoaugmentation technology for phytoremediation of environmental pollutants: current scenario and future prospects. In: Bioremediation for environmental sustainability 2021 Jan 1, pp 329–381. Elsevier

Kumari K, Cherian S, Bauddh K (2022) Microbial augmented phytoremediation with improved ecosystems services. InAdvances in microbe-assisted phytoremediation of polluted sites 2022 Jan 1, pp 27–62. Elsevier

Lim MW, Von Lau E, Poh PE (2016) A comprehensive guide of remediation technologies for oil contaminated soil—Present works and future directions. Mar Pollut Bull 109(1):14–45

Lu XY, Zhang T, Fang HH (2011) Bacteria-mediated PAH degradation in soil and sediment. Appl Microbiol Biotechnol 89:1357–1371

Macaulay BM (2015) Understanding the behaviour of oil-degrading micro-organisms to enhance the microbial remediation of spilled petroleum. Appl Ecol Environ Res 13(1):247–262

Maletić SP, Beljin JM, Rončević SD, Grgić MG, Dalmacija BD (2019) State of the art and future challenges for polycyclic aromatic hydrocarbons is sediments: sources, fate, bioavailability and remediation techniques. J Hazard Mater 5(365):467–482

Naik PR, Sakthivel N. Functional characterization of a novel hydrocarbonoclastic Pseudomonas sp. strain PUP6 with plant-growth-promoting traits and antifungal potential. Res Microbiol 157(6):538–46

Philp JC, Whiteley AS, Ciric L, Bailey MJ. Monitoring bioremediation. Bioremediation: applied microbial solutions for real-world environmental cleanup 29:237–68

Rodrigues DF, Sakata SK, Comasseto JV, Bícego MC, Pellizari VH (2009) Diversity of hydrocarbon-degrading Klebsiella strains isolated from hydrocarbon-contaminated estuaries. J Appl Microbiol 106(4):1304–1314

Rojo F (2010) Enzymes for aerobic degradation of alkanes. Handbook Hydrocarbon Lipid Microbiol 2:781–797

Salminen JM, Tuomi PM, Jørgensen KS (2008) Functional gene abundances (*nahAc, alkB, xylE*) in the assessment of the efficacy of bioremediation. Appl Biochem Biotechnol 151:638–652

Sierra-Garcia IN, de Oliveira VM (2013) Microbial hydrocarbon degradation: efforts to understand biodegradation in petroleum reservoirs. Biodegradation-Eng Technol 14(10):55920

Sihag S, Pathak H, Jaroli DP (2014) International journal of pure & applied bioscience. Int J Pure App Biosci 2(3):185–202

Sri Ranjan R, Qian Y, Krishnapillai M (2006) Effects of electrokinetics and cationic surfactant cetyltrimethylammonium bromide [CTAB] on the hydrocarbon removal and retention from contaminated soils. Environ Technol 27(7):767–776

Tang J, Zhu W, Kookana R, Katayama A (2013) Characteristics of biochar and its application in remediation of contaminated soil. J Biosci Bioeng 116(6):653–659

Throne-Holst M, Markussen S, Winnberg A, Ellingsen TE, Kotlar HK, Zotchev SB (2006) Utilization of n-alkanes by a newly isolated strain of Acinetobacter venetianus: the role of two AlkB-type alkane hydroxylases. Appl Microbiol Biotechnol 72:353–360

Tortella G, Durán N, Rubilar O, Parada M, Diez MC (2015) Are white-rot fungi a real biotechnological option for the improvement of environmental health? Crit Rev Biotechnol 35(2):165–172

Van Elsas JD, Trevors JT, Jansson JK, Nannipieri P (2006) Modern soil microbiology. CRC Press

Wang L, Tang Y, Wang S, Liu RL, Liu MZ, Zhang Y, Liang FL, Feng L (2006) Isolation and characterization of a novel thermophilic Bacillus strain degrading long-chain n-alkanes. Extremophiles 10:347–356

Widdel F, Grundmann O (2010) Biochemistry of the anaerobic degradation of non-methane alkanes. InHandbook of hydrocarbon and lipid microbiology

Wrenn BA, Venosa AD (1996) Selective enumeration of aromatic and aliphatic hydrocarbon degrading bacteria by a most-probable-number procedure. Can J Microbiol 42(3):252–258

Wu M, Ye X, Chen K, Li W, Yuan J, Jiang X (2017) Bacterial community shift and hydrocarbon transformation during bioremediation of short-term petroleum-contaminated soil. Environ Pollut 1(223):657–664

Xu X, Liu W, Tian S, Wang W, Qi Q, Jiang P, Gao X, Li F, Li H, Yu H (2018) Petroleum hydrocarbon-degrading bacteria for the remediation of oil pollution under aerobic conditions: a perspective analysis. Front Microbiol 3(9):2885

Yesilada O, Birhanli E, Geckil H (2018) Bioremediation and decolorization of textile dyes by white rot fungi and laccase enzymes. Mycoremediation Environ Sustain 2:121–153

Yuste L, Corbella ME, Turiégano MJ, Karlson U, Puyet A, Rojo F (2000) Characterization of bacterial strains able to grow on high molecular mass residues from crude oil processing. FEMS Microbiol Ecol 32(1):69–75

Zelles L (1999) Fatty acid patterns of phospholipids and lipopolysaccharides in the characterisation of microbial communities in soil: a review. Biol Fertil Soils 29:111–129

Chapter 11
Advanced Environmental Nanotechnology for the Degradation Waxes, Asphaltenes and Paraffins in Oil and Gas Industries

Akshobya Kumar, Subhasis Patra, Sankha Chakrabortty, and Suraj Kumar Tripathy

Abstract Wax, asphaltene, and paraffin degradation in the oil and gas sectors offers a wide range of challenges that can be addressed using advanced environmental nanotechnology. It often leads to pipeline obstructions, operational inefficiencies, and higher maintenance expenses. Traditional methods for their removal and remediation are often costly, environmentally damaging, and inefficient. This review represents an overview of nanotechnology based solutions for the degradation of waxes, asphaltenes, and paraffins. Nanomaterial provides unique properties such as high surface area-to-volume ratios, increased reactivity, optical activity, thermo physical and mechanical properties. Due to certain characteristics of nanomaterial hydrocarbon deposits can be effectively removed from oil and gas industries by means of efficient adsorption, dispersion, and catalytic degradation.

11.1 Introduction

11.1.1 Overview of the Challenges Posed by Waxes, Asphaltenes, and Paraffins in Oil and Gas Industries

Wax deposition is one of the biggest operational challenges in oil and gas sector. Crude oil transportation via pipelines is quite difficult, mostly because of temperature changes along the pipeline's path (Olajire 2021). The crude oil's composition and pour point frequently alter as a result of these temperature fluctuations. As a result, the

A. Kumar
School of Biotechnology, Kalinga Institute of Industrial Technology, Bhubaneswar, India

S. Patra · S. Chakrabortty (✉) · S. K. Tripathy
School of Chemical Engineering, Kalinga Institute of Industrial Technology, Bhubaneswar, India
e-mail: sankha.chakrabortty@kiitbiotech.ac.in

S. Banerjee et al. (eds.), *Environmental Friendly Green Technologies for Improvement of Heavy Crude Oil Flow Assurance*, Environmental Science and Engineering,
https://doi.org/10.1007/978-3-031-86701-9_11

accumulation of wax and asphaltenes happens, which causes the pipelines' diameter to decrease and causes obstructions. The formation of damage, flow restriction, and, in extreme situations, the abandonment of wells and production system flow lines are all significant consequences of these deposits. With historical references going back to 1928, wax buildup has been acknowledged as a persistent issue that emerges throughout the production, transportation, and storage of oil. There are problems with wax buildup in flow lines, surface equipment, topside facilities, and downstream refineries, among other places in the petroleum supply chain (Anisuzzaman and Karali 2023; Elarbe et al. 2021).

11.1.2 Importance of Addressing the Challenges for Environmental Protection

Crude oil's original rheology can be significantly changed by wax precipitation and deposition. The crude oil becomes more viscous when the wax crystals precipitate and deposit, turning it into a highly viscous liquid that sticks to the inside walls of pipes. As a result, the pipe's effective cross-sectional area decreases, obstructing oil flow and creating operating difficulties (Mohyaldinn et al. 2019). The buildup of wax and asphaltene deposits frequently results in high maintenance and cleaning costs for machinery such as refineries, pumps, and pipelines. There can be serious environmental repercussions from improper management of wax and asphaltene deposition. The environment can be contaminated by spills and leaks resulting from equipment malfunctions or pipeline blockages, endangering human health, wildlife, and ecosystems (Sousa et al. 2022).

11.1.3 To Explore Advanced Nanotechnological Approaches for Their Degradation

In recent years, there has been a great deal of interest in the study of modern nanotechnological methods for the breakdown of paraffin, wax, and asphaltene in the oil and gas sector. Since nanotechnology can create and engineer materials at the nanoscale level to provide enhanced qualities and capabilities, it presents a promising avenue for solving problems (Farooq et al. 2021). In the oil and gas sector, nanotechnologies are revolutionizing flow assurance, especially in the battle against solid deposits that hinder pipeline operations. The damage caused by paraffin, wax, and asphaltene deposition can be prevented by using nanoparticles. Utilising Co-Ni nanoparticles and a unique polymer, researchers have created a novel technique to stop paraffin from depositing. These nanoparticles can be functionalized to specifically target deposits, making it easier to break them down and remove them from surfaces (Yang et al. 2015). Several silica nanoparticles were investigated, and the most successful ones were used to create nanofluid. This nanofluid's injection decreased formation damage caused by asphaltene, increasing oil recovery by 11% (Guerrero-Martin et al. 2023).

11.2 Waxes, Asphaltenes, and Paraffins: Characteristics and Challenges

Crude oils contain essential components called waxes and asphaltenes that have a big impact on their qualities (Ganeeva et al. 2016). Waxes are a mixture of long-chain hydrocarbons, typically containing straight-chain alkanes, branched alkanes, and cyclic compounds. They are solid at room temperature and, in some situations, can precipitate out of crude oil and leave deposits on the surfaces of equipment. Paraffin waxes are composed of normal alkanes (n-alkanes) with a chain length ranging from C18 to C30. These waxes crystallize into large flat plates(macrocrystalline) with a melting point typically falling between 40 and 60 °C. On the other hand, microcrystalline waxes are defined by their composition primarily comprising higher molecular weight n-alkanes within the C30–C60 range. They possess a notable presence of iso-paraffins and naphthenes, leading to significantly wider melting point ranges, often exceeding 60–90 °C (Freund et al. 1982). High molecular weight molecules called asphaltenes are made up of aliphatic and aromatic hydrocarbons and heteroatoms like oxygen, sulphur, and nitrogen. They are soluble in aromatics like toluene but insoluble in light n-alkanes like n-pentane or n-heptane (Fakher et al. 2020; Thota and Onyeanuna 2016). The deposition of these compounds on pipeline surfaces can lead to flow restrictions, increased pressure drop, and reduced flow rates.

11.3 Traditional Methods for Degradation

Conventional techniques employed for the degradation of waxes, asphaltenes, and paraffins reveals a range of established methods utilized across the oil and gas industry. These methods typically involve thermal treatments, chemical solvents, mechanical agitation.

11.3.1 Thermal Methods

Since wax deposition is temperature-sensitive, wax deposits inside reservoirs can be effectively prevented and remedied by using thermal procedures. The following lists several procedures:-

11.3.1.1 Hot Oiling

The frequently used technique of "hot oiling" reduces clogs in wellbores and tubing by melting and removing wax buildup in flow lines and downhole. In order to soften waxes blocking flow in tubing or fittings, this approach involves heating the oil above the melting point of the wax and then injecting the heated oil downhole (Thota and Onyeanuna 2016). However, a notable drawback of hot oiling is the evaporation of lighter elements, impacting the overall value of the oil during the dewaxing process. Moreover, there is a risk of damaging permeability if melted wax comes into the structure (Elarbe et al. 2021).

11.3.1.2 Direct Heating

When dealing with wax-related flow problems in pipes, direct heating is thought to be quite beneficial. It gives important benefits by regulating temperatures above the formation region. To produce heat, a significant electric current is run through the pipeline wall. This method is especially reliable and appealing for Deepwater field activities where flowlines need to be transported. The high expense of heating system maintenance and the lack of a source for producing power are disadvantages, too (Elarbe et al. 2021).

11.3.2 Mechanical Method

The most common technique for eliminating wax accumulation, particularly in offshore subsea systems, is pigging. Pigs are made to push on the pipe wall with a mechanical force and clear debris through tiny bypass holes. The type of pig utilised depends on how severe the deposits are. Pigging activities can cause production halts and financial losses even though they have low operational costs. Pigging is effective, but if it is done too often, it can potentially harm the formation and produce problems with attachment (Elarbe et al. 2021) (Fig. 11.1).

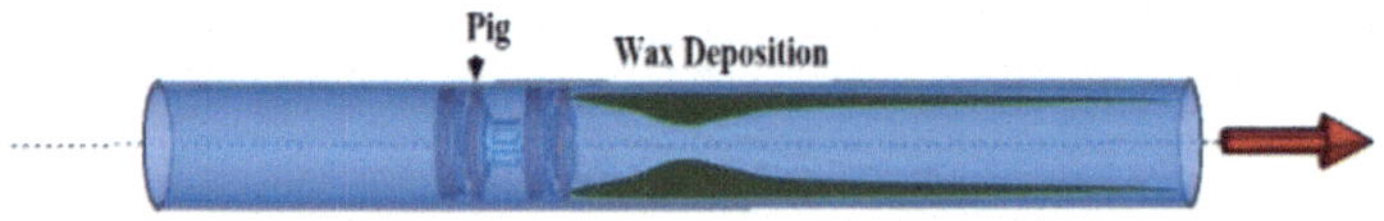

Fig. 11.1 Pigging process for removing the deposit from the pipeline (Elarbe et al. 2021)

11.3.3 Chemical Method

The utilization of a chemical wax inhibitor can significantly decrease the rate of wax deposition in certain crude oil production systems, presenting an economically attractive option to enhance output or reduce wax remediation expenses. Solvent treatments are highly effective but costly methods for addressing wax and asphaltene depositions, typically used when hot oil or hot water techniques fail (Thota and Onyeanuna 2016).

11.4 Nanotechnology: Fundamentals and Application in Oil and Gas Industry

11.4.1 Introduction to Nanotechnology and Its Principles

Nanotechnology is a multidisciplinary field of science and engineering focused on manipulating matter at the nanoscale, typically ranging from 1 to 100 nm (Bayda et al. 2019). Across a wide range of scientific and engineering fields, the study and application of materials at nanoscale dimensions (1–100 nm) has become one of the fastest growing fields. Nanotechnology has significantly altered the perspectives of modern scientists and engineers on smart materials. Advancements in nanotechnology have enabled scientists to create a wide range of unique and cutting-edge nanomaterials, nanodevices, and nanotools that can be used in a variety of industries, including electronics, aerospace, medicine, biomedical fields, smart materials and manufacturing, pharmaceuticals, photography, energy, and more (Malik et al. 2023). The use of nanoparticles in modern life has increased due to their special qualities. Nanotechnology offers transformative solutions for achieving environmental sustainability by addressing pollution, water scarcity, and energy challenges through innovative materials and technologies. The unique properties of nanomaterials enable enhanced performance and efficiency in environmental applications, paving the way for a cleaner, greener future (Fig. 11.2).

11.4.2 Nanotechnology in Oil and Gas Industry

The oil and gas sector has played a significant role in this transition and is currently investigating the advantages of this technology. Many nanomaterials have been widely used in various oil and gas sector operations, such as magnetic nanoparticles, metallic nanoparticles, carbon nanotubes, and metal oxide nanoparticles. Nanoparticles can be synthesized using diverse methods, with the sol–gel and hydrothermal processes standing out as efficient techniques for producing crystalline nanoparticles (Franco et al. 2021). It plays different role in oil industries

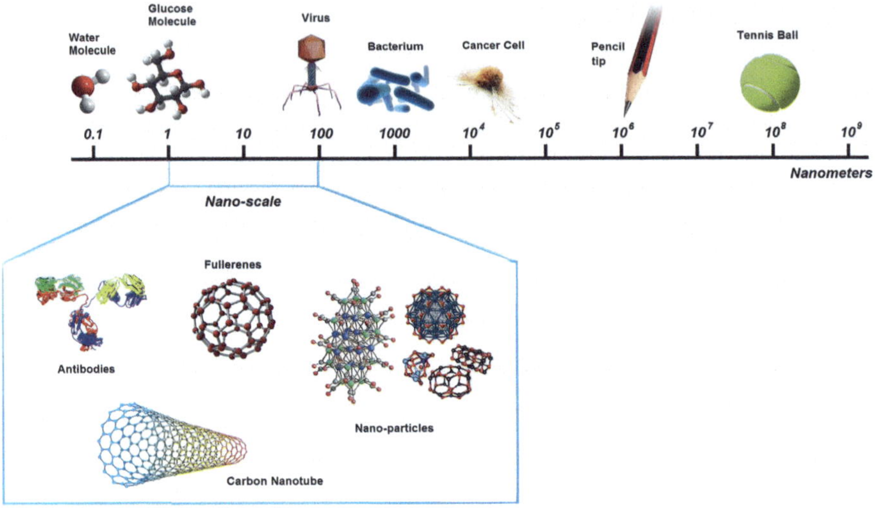

Fig. 11.2 A length scale comparison (Lau et al. 2017)

such as enhanced oil recovery, applications in drilling, counteract several methods of formation damage, such as the precipitation and deposition of asphaltene, improved oil recovery and in cementing (Franco et al. 2021). Nanoparticles exhibit unique physicochemical properties that make them attractive candidates for improving oil recovery efficiency. Engineered nanoparticles, including silica, carbon nanotubes, and metal oxides, can be functionalized to target specific reservoir conditions and interactions. These nanoparticles interact with reservoir fluids, rock surfaces, and oil–water interfaces, influencing key EOR mechanisms such as wettability alteration, interfacial tension reduction, and mobility control. By tailoring nanoparticle properties, researchers can optimize their performance in various reservoir environments and operational scenarios (Agista et al. 2018; Oladiji et al. Nutrition and diet in health; Cheraghian and Hendraningrat 2016; Sun et al. 2017). Nanomaterials, including nanoparticles, nanotubes, and nanocomposites, have gained attention for their potential applications in reservoir characterization. Functionalized nanoparticles can serve as tracers for monitoring fluid flow and migration within the reservoir, providing valuable information on reservoir connectivity and compartmentalization. Moreover, nanocomposites embedded with nanoparticles can enhance the mechanical and thermal properties of wellbore cement and drilling fluids, improving well integrity and reservoir stability. Additionally, nanoscale additives can modify the wettability of reservoir rocks, influencing fluid behavior and recovery efficiency (Iravani et al. 2023). Nanostructured coatings represent a promising avenue for corrosion protection, providing enhanced barrier properties, adhesion, and durability compared to conventional coatings (Nazari et al. 2022). Various types of nanocoatings, including nanocomposites, nanoclays, and graphene-based coatings, have been developed to inhibit corrosion initiation and propagation on metal surfaces (Kausar et al. 2022). It offer targeted protection against corrosion by forming a protective barrier or altering the electrochemical environment at the metal surface. Examples include cerium oxide nanoparticles, silver nanoparticles, and polymer-encapsulated inhibitors, which exhibit prolonged inhibition efficiency and reduced environmental toxicity (Muresan 2023).

Surface engineering techniques utilizing nanotechnology enable the creation of tailored surface morphologies and functionalities to enhance corrosion resistance (Ramezani et al. 2023). Nanostructured surfaces, such as nanotubes, nanowires, and nanoporous coatings, can modulate surface energy, wettability, and adhesion, thereby minimizing corrosion susceptibility and promoting self-healing properties (Farooq et al. 2022). The integration of nanotechnology into drilling fluids represents a promising avenue for advancing drilling technology and optimizing well construction processes (Alkalbani and Chala 2024). By leveraging nanomaterials to tailor drilling fluid properties with precision, operators can enhance drilling efficiency, minimize environmental impact, and maximize hydrocarbon recovery it aims at addressing technical challenges, optimizing nanoparticle formulations, and assessing environmental implications will be critical for realizing the full potential of nanotechnology in the oil and gas industry (Rafati et al. 2018).

Overall Nanotechnology holds great promise for revolutionizing reservoir characterization practices in the oil and gas industry by offering enhanced spatial resolution, sensitivity, and functionality at the nanoscale. Through the utilization of nanomaterials, nanosensors, and nanoscale imaging techniques, researchers and practitioners can gain deeper insights into reservoir properties, fluid behavior, and pore-scale processes, ultimately leading to more accurate reservoir models and optimized hydrocarbon recovery strategies (Zhe and Yuxiu 2018) (Fig. 11.3).

Nanoparticles offer a promising avenue for addressing deposit buildup in pipelines. These nanoparticles can be engineered to adsorb onto wax, paraffin, and asphaltene molecules, disrupting their aggregation and promoting their dispersion within the pipeline fluid. Moreover, nanoparticles with catalytic properties can facilitate the oxidation or decomposition of these deposits, breaking them down into smaller, manageable fragments that can be easily transported through the pipeline. In addition to nanoparticles, nanocomposite coatings represent another effective strategy for mitigating deposit formation in pipelines. These coatings, applied to the inner surface of pipelines, can prevent the adhesion of wax, paraffin, and asphaltene deposits, thereby minimizing the risk of flow restrictions and blockages. Nanoparticles possess a significant specific surface area per unit volume, which enhances their adsorption and catalytic characteristics (Roy et al. 2023). Numerous nanoparticles, including silica (Al-Shboul et al. 2023), clay (Ning et al. 2022), and carbon nanostructures (Zhang et al. 2021; Mohammadi and Mahmoudi Alemi 2021), have been altered and employed as wax inhibitors. Based on wax interaction with inhibitor there are three stages namely nucleation, co-crystallization and adsorption. Wax crystals have several nucleation sites created by the nucleation inhibition mechanism, which inhibits percolation and shifts the thermodynamic equilibrium results in weakening the wax structure (Roy et al. 2023). The majority of inhibitors based on nanoparticles are discovered as templates for the nucleation of wax crystals, including carbon

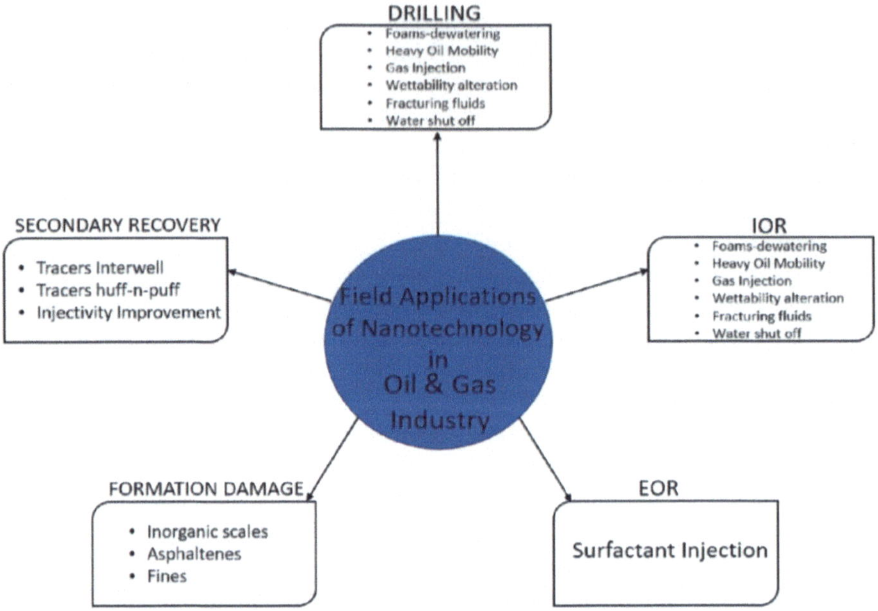

Fig. 11.3 Various uses of nanotechnology in oil and gas industries (Juyal et al. 2023)

nano hybrid (Mohammadi and Mahmoudi Alemi 2021), clay (Ning et al. 2022), and silica (Al-Shboul et al. 2023). Another process involves forming a ferrofluid in the annulus between the production tube and casing by suspending Co–Ni nanoparticles in a carrier fluid. It is carried out when temperature is low enough for crude oil to precipitate wax. Toroid-shaped inductors are positioned and nanoparticles are heated by the alternating magnetic field these inductors apply across them which results in melting of wax deposit (Haindade et al. 2012). Another way to deal with deposition of hydrocarbons is using of supermagnetic Fe3O4 which has been embedded in thin layer of cured epoxy termed 'nanopaint. The coating in the internal surface of the flow lines generate heat which prevents formation of wax deposition (Mehta et al. 2014). The deposition of asphalatene can be avoided by adding magnetic nanoparticle. Due to self associative property of asphalatene they form aggregates. These aggregates formation can be restrained by adding magnetic nanoparticle which leads to more interaction between nanoparticle and asphalatene hence, no deposition of asphalatene (Zhou et al. 2020). Asphalatene can also be adsorbed by nanoparticles. There have been four types of NPs taken into consideration: carbonaceous nanostructures, pure metal oxides, hybrid NPs, and surface-coated NPs. For the purpose of overcoming the limitations of single NPs and achieving new customised qualities, hybrid NPs are generated by merging at least two distinct NPs. Each individual NP can be coated with a second ceramic or polymer layer (such as polythiophene (PT)) to create the coated NPs (Fig. 11.4).

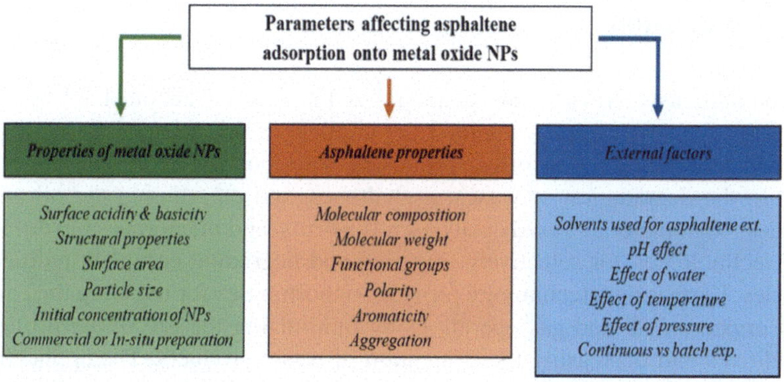

Fig. 11.4 Factors that influence asphaltene's adsorption onto metal oxide nanoparticles (Farooq et al. 2021)

Moreover, adsorption decreased with increasing temperature and increased with increasing pressure. When the initial asphaltene concentration is raised and the NP content is lowered, adsorption is also improved. Nanotechnological approach overcomes the limitation offered by traditional methods. Detailed comparison is shown in Table 11.1.

Graphene-based nanocomposites, for example, exhibit superior filtration properties due to their high surface area and tunable pore sizes. By incorporating nanocomposites into drilling fluids it can also reduce formation damage, and improve well productivity. Nanofluids, suspensions of nanoparticles in a base fluid, have also shown promise in mitigating formation damage caused by fines migration and clay swelling.

Table 11.1 Comparison of traditional method with nanotechnological approach for degrading hydrocarbons

Aspect	Traditional method	Nanotechnological approach
Efficiency	May vary based on conditions	Generally higher efficiency due to targeted approach and increased surface area for reaction
Precision	Limited precision	High precision targeting specific hydrocarbons, minimizing collateral damage
Environmental impact	May involve harsh chemicals	Often more eco-friendly, reducing the use of hazardous chemicals and byproducts
Maintenance	Regular maintenance required	Requires less frequent maintenance due to longer-lasting effects of nanomaterials

11.5 Conclusion

This comprehensive review has emphasized the role of nanotechnology in the revolution of oil fields and examined the challenges related to flow in pipelines in oil and gas industries. It offers promising solutions for addressing challenges related to the degradation of waxes, asphaltenes, and paraffins in the oil and gas industries. Nanomaterial-based catalysts, adsorbents, and membranes have emerged as key technologies for selectively targeting and degrading complex hydrocarbon molecules. Further nanotechnology provides opportunities for reducing the environmental impact of oil and gas operations by minimizing the use of harmful chemical additives and decreasing the generation of waste products. The application of nanotechnology to the oil and gas industry has the potential to make hydrocarbon recovery and processing more economical and resource-efficient. Overall the oil and gas industries have a lot to gain from the use of advanced environmental nanotechnology in terms of improved operational efficiency, reduced environmental risk, and the advancement of more sustainable energy methods.

References

Agista MN, Guo K, Yu Z (2018) A state-of-the-art review of nanoparticles application in petroleum with a focus on enhanced oil recovery. Appl Sci 8(6):871

Alkalbani AM, Chala GT (2024) A comprehensive review of nanotechnology applications in oil and gas well drilling operations. Energies 17(4):798

Al-Shboul T, Sagala F, Nassar NN (2023) Role of surfactants, polymer, nanoparticles, and its combination in inhibition of wax deposition and precipitation: a review. Adv Colloid Interface Sci 102904

Anisuzzaman SM, Karali R (2023) A short review on waxes and asphaltenes deposition and various treatment approaches to mitigate deposition problems in crude oil. In: AIP conference proceedings, vol 2945, no 1. AIP Publishing

Bayda S, Adeel M, Tuccinardi T, Cordani M, Rizzolio F (2019) The history of nanoscience and nanotechnology: from chemical–physical applications to nanomedicine. Molecules 25(1):112

Cheraghian G, Hendraningrat L (2016) A review on applications of nanotechnology in the enhanced oil recovery part a: effects of nanoparticles on interfacial tension. Int Nano Lett 6:129–138

Elarbe B, Elganidi I, Ridzuan N, Abdullah N, Yusoh K (2021) Paraffin wax deposition and its remediation methods on crude oil pipelines: a systematic review. Maejo Int J Energy Environ Commun 3(3):6–34

Fakher S, Ahdaya M, Elturki M, Imqam A (2020) Critical review of asphaltene properties and factors impacting its stability in crude oil. J Petrol Explor Prod Technol 10:1183–1200

Farooq U, Patil A, Panjwani B, Simonsen G (2021) Review on application of nanotechnology for asphaltene adsorption, crude oil demulsification, and produced water treatment. Energy Fuels 35(23):19191–19210

Farooq SA, Raina A, Mohan S, Arvind Singh R, Jayalakshmi S, Irfan Ul Haq M (2022) Nanostructured coatings: review on processing techniques, corrosion behaviour and tribological performance. Nanomaterials 12(8):1323

Franco CA, Franco CA, Zabala RD, Bahamón I, Forero A, Cortés FB (2021) Field applications of nanotechnology in the oil and gas industry: recent advances and perspectives. Energy Fuels 35(23):19266–19287

Freund M, Csikós R, Keszthelyi S, Mózes G (1982) I. Chemical, crystallographical and physical properties of liquid paraffins and paraffin waxes. Dev Pet Sci 14:13–140

Ganeeva YM, Yusupova TN, Romanov GV (2016) Waxes in asphaltenes of crude oils and wax deposits. Pet Sci 13:737–745

Guerrero-Martin CA, Montes-Pinzon D, Motta M, da Silva M, Montes-Paez E, Guerrero-Martin LE, Salinas-Silva R, Szklo A (2023) Asphaltene precipitation/deposition estimation and inhibition through nanotechnology: a comprehensive review. Energies 16(13):4859

Haindade Z, Bihani AD, Javeri S, Jere C (2012) Enhancing flow assurance using Co-Ni nanoparticles for dewaxing of production tubing. In: SPE international oilfield nanotechnology conference and exhibition. SPE, pp SPE-157119

Iravani M, Khalilnezhad Z, Khalilnezhad A (2023) A review on application of nanoparticles for EOR purposes: history and current challenges. J Pet Explor Prod Technol 13(4):959–994

Kausar A, Ahmad I, Bocchetta P (2022) High-performance corrosion-resistant polymer/graphene nanomaterials for biomedical relevance. J Compos Sci 6(12):362

Malik S, Muhammad K, Waheed Y (2023) Nanotechnology: a revolution in modern industry. Molecules 28(2):661

Mehta P, Huh C, Bryant SL (2014) Evaluation of superparamagnetic nanoparticle-based heating for flow assurance in subsea flowlines. In: International petroleum technology conference. IPTC, pp IPTC-18090

Mohammadi S, Mahmoudi Alemi F (2021) Simultaneous control of formation and growth of asphaltene solids and wax crystals using single-walled carbon nanotubes: an experimental study under real oilfield conditions. Energy Fuels 35(18):14709–14724

Mohyaldinn ME, Husin H, Hasan N, Elmubarak MM, Genefid AM, Dheeb ME (2019) Challenges during operation and shutdown of waxy crude pipelines. Process Heavy Crude Oils—Challenges Oppor IntechOpen: Rijeka, Croatia, 153

Muresan LM (2023) Nanocomposite coatings for anti-corrosion properties of metallic substrates. Materials 16(14):5092

Nazari MH, Zhang Y, Mahmoodi A, Xu G, Yu J, Wu J, Shi X (2022) Nanocomposite organic coatings for corrosion protection of metals: A review of recent advances. Prog Org Coat 162:106573

Ning X, Song X, Zhang S, Wang Y, Feng Y (2022) Insights into flow improving for waxy crude oil doped with EVA/SiO2 nanohybrids. ACS Omega 7(7):5853–5863

Oladiji AT, Oladele JO, Ajayi EI Nutrition and diet in health

Olajire AA (2021) Review of wax deposition in subsea oil pipeline systems and mitigation technologies in the petroleum industry. Chem Eng J Adv 6:100104

Rafati R, Smith SR, Haddad AS, Novara R, Hamidi H (2018) Effect of nanoparticles on the modifications of drilling fluids properties: a review of recent advances. J Petrol Sci Eng 161:61–76

Ramezani M, Mohd Ripin Z, Pasang T, Jiang CP (2023) Surface engineering of metals: techniques, characterizations and applications. Metals 13(7):1299

Roy A, Kabra D, Pareek G, Kumari K, Kashyap PP, Naik S, Selvaraj SK (2023) A review of nanomaterials and their applications in oil and petroleum industries. Nano Express

Sousa AM, Ribeiro TP, Pereira MJ, Matos HA (2022) Review of the economic and environmental impacts of producing waxy crude oils. Energies 16(1):120

Sun X, Zhang Y, Chen G, Gai Z (2017) Application of nanoparticles in enhanced oil recovery: a critical review of recent progress. Energies 10(3):345

Thota ST, Onyeanuna CC (2016) Mitigation of wax in oil pipelines. Int J Eng Res Rev 4(4):39–47

Yang J, Ji S, Li R, Qin W, Lu Y (2015) Advances of nanotechnologies in oil and gas industries. Energy Explor Exploit 33(5):639–657

Zhang J, Yu H, Liang Y, Sun Z, Liu Y, Jing G (2021) Influence of EVA/nano-sepiolite on the wax crystal and rheological property of Daqing crude oil. J Dispers Sci Technol 42(8):1125–1131

Zhe Z, Yuxiu A (2018) Nanotechnology for the oil and gas industry—an overview of recent progress. Nanotechnol Rev 7(4):341–353

Zhou K, Zhou X, Liu J, Huang Z (2020) Application of magnetic nanoparticles in petroleum industry: a review. J Petrol Sci Eng 188:106943

Chapter 12
Applications of Non-pathogenic Fungi in Enhanced Oil Recovery

Neha Saxena, Merajul Islam, Prathibha Pillai, and Amit Kumar

Abstract As the demand for oil and its byproducts rises across the world, the supply of traditional crude oil is steadily dwindling. In order to keep up with energy demands, oil corporations are pumping out crude oil from older fields, where traditional primary and secondary recovery has left more than 60% of reserves untapped (Zhang et al. Crit Rev Biotechnol 40:459–474, 2020; Shibulal et al. Sci World J 2014:1–12, 2014). To raise the total productivity of oil from matured oil reservoirs, it is necessary to create an inexpensive and effective enhanced oil recovery (EOR) technique, such as chemical, thermal, microbiological, or miscible EOR, to generate this residual oil. One of the techniques used in the tertiary recovery stage is called Microbial Enhanced Oil Recovery (MEOR). Heavy oil is notoriously difficult to remove from oil reservoirs due to its high viscosity and limited flow characteristics. Microbial EOR is the new emerging area in the field of EOR as it combines green chemistry with lower carbon footprint. The nonpathogenic fungi like mycorrhiza, Trichoderma sp., Penicillium sp., Fusarium sp., Phoma sp. etc. have been identified as an source of rhamnolipids, Sophorolipids that produces the various natural surfactants commonly called "Biosurfactants". By decreasing the interfacial tension between the oil and water, a biosurfactant solution can free the oil that has been trapped by capillary forces interface and hence found application in enhanced oil recovery. In this chapter we have focused on the application of various non pathogenic fungi in enhanced oil recovery.

Keywords Biosurfactant · Non pathogenic fungi · Interfacial tension · Enhanced oil recovery

N. Saxena (✉) · M. Islam
School of Basic Sciences and Technology, IIMT University, Meerut, Uttar Pradesh, India
e-mail: gemini.neha1990@gmail.com

P. Pillai
Halliburton, Bengaluru, India

A. Kumar
Department of Petroleum Engineering and Geoengineering, Rajiv Gandhi Institute of Petroleum Technology, Jais, Uttar Pradesh, India

S. Banerjee et al. (eds.), *Environmental Friendly Green Technologies for Improvement of Heavy Crude Oil Flow Assurance*, Environmental Science and Engineering,
https://doi.org/10.1007/978-3-031-86701-9_12

12.1 Introduction

The anticipated rise in global energy consumption is attributed to the projected growth in population, urbanisation, industrialization, and economic advancement. The economic condition of the majority of nations relies heavily on petroleum products and its derivatives (Zhang et al. 2020). India, in the year 2019, exhibited substantial energy consumption, amounting to about 806.1 million metric tonnes of petroleum comparable. Within this total, crude petroleum consumption accounted for 254.39 million metric tonnes. These figures indicate that India's energy consumption constituted 5.8% of the global main source of energy usage during the specified period. As of March 31, 2020, the crude oil reserves in India amounted to around 603.36 million metric tonnes, with a production contribution of approximately 32.17 million metric tonnes (Shibulal et al. 2014). In order to fulfil the growing energy requirements and address the limitations on available assets, oil business owners will need to engage in the exploitation of more intricate reservoirs. However, in order to achieve this objective, the utilisation of very sophisticated technology will be necessary. Hence, the extraction of oil deposits, regardless of their developmental stage (undeveloped, mature, or yet to be discovered), necessitates the use of Enhanced Oil Recovery (EOR) methods. Traditional primary and secondary methods are insufficient in effectively extracting the residual oil present in capillary rock pores, especially those with a diameter less than 10 nm. The challenge of oil recovery arises from several factors, including the decrease in pressure inside oil reservoirs over time due to oil generation, petroleum product exhibits a high level of viscosity, and the subsequent reduction in rates of movement (Safdel et al. 2017; Nikolova and Gutierrez 2020; She et al. 2019; Rivera et al. 2019; Romero-Zeron 2016; Bi et al. 2019).

The major goal of processing facilities of gasoline is to recover leftover lubricant after traditional procedures have been employed. This residual oil accounts for roughly 5 trillion barrels, which is equivalent to 70% of the global oil reserves (Zhang et al. 2020). Numerous firms, facing financial constraints in the recovery of residual oil, are presently engaged in the pursuit of cost-effective and effective machinery for the extraction of traditional petroleum. A significant portion, ranging from 35 to 55%, of the remaining petroleum stored in tanks remains untapped following traditional recovery methods. Consequently, tertiary techniques, also known as Enhanced Oil Recovery (EOR), are employed to augment the output of crude oil (Shibulal et al. 2014; Safdel et al. 2017).

The conversion of these applications into projects is contingent upon their alignment with the economic restrictions of the prevailing global oil marketplace, so effectively prolonging the operational lifespan of the wells. Furthermore, these approaches are environmentally benign and serve as a viable substitute for conventional enhanced oil recovery (EOR) methods. The approach of microbial enhanced oil recovery (MEOR) integrates the unique qualities of microorganisms with conventional petroleum engineering techniques, resulting in the development of a novel bioengineering technology. The aforementioned methodology ultimately optimises the amount of oil reserves that can be extracted, prolongs the operational lifespan of oil fields, and enhances the recovery factor of individual wells (She et al. 2019; Rivera et al. 2019). The promotion of microbial activity by MEOR facilitates the release of metabolic products by the bacteria, which include biomass, biopolymers, biosurfactants, biosolvents and bioacids (Nikolova and Gutierrez 2020). Every single one of these metabolic offspring has a distinct role in the microbial system and enhances oil recovery on an individual basis. The advantageous aspect of these microorganisms is in their capacity to increase the fluidity of petroleum by making it less thick. The aforementioned characteristic is effectively utilised by supplying the microorganisms with appropriate nourishment to generate a satisfactory amount of metabolic by-products and by prolonging the lifespan of the well (She et al. 2019; Rivera et al. 2019).

Microorganisms have the potential to synthesise valuable, environmentally degradable, and comparatively less harmful substances through the utilisation of cost-effective substrates or raw resources. The adoption of microbial enhanced oil recovery (MEOR) as a feasible substitute for chemically enhanced oil recovery (CEOR) has gained attention due to the perception that CEOR is a financially burdensome technology with limited environmental sustainability. Microbial Enhanced Oil Recovery (MEOR) involves the use of certain microbial strains to produce chemicals similar to those employed in conventional Chemical Enhanced Oil Recovery (CEOR) methods, with the aim of enhancing the extraction of petroleum extracted from shallow or depleted sources. Emergence of microbial technologies recognised globally as profitable and environmentally viable approaches for enhancing oil production (Rivera et al. 2019; Romero-Zeron 2016; Bi et al. 2019).

This chapter offers a comprehensive examination of the many steps involved in oil recovery, with a specific emphasis on the microbial approach. This technique is highlighted because to the wide-ranging advantages offered by the metabolic by-products it produces, namely biosurfactants. This paper is a summary of several research that have evaluated the potential and economic viability of biosurfactants for MEOR. The authors further expound upon the essential principles of MEOR, in addition to discussing its potential applications and the obstacles encountered. Moreover, this research examines the historical and contemporary landscape of Microbial Enhanced Oil Recovery (MEOR), along with its underlying dynamics, while also presenting a concise overview of the mechanism employed by hydrocarbon-degrading microorganisms. After carefully reviewing a number of case studies conducted globally, encompassing both laboratory and field experiments, this study aims to evaluate the historical efficacy and future prospects of MEOR techniques. This response provides a concise overview of the current developments and endeavours pertaining to MEOR. This research further elucidates possible future prospects for the advancement and use of MEOR, while emphasising the significance of a unified methodology in computer-aided design and modelling.

12.2 Phases of Oil Recovery

The process of extracting oil is often divided into three distinct levels, specifically the initial, intermediate, and final phases. Traditional primary and secondary techniques employed in the extraction of oil deplete a significant portion, ranging from less than a quarter to half, amount of oil in storage within a well in a very short period of time. The issue of achieving a significant degree of extraction is tackled through the use of an advanced method known as EOR (see Fig. 12.1) (Haq 1946).

12.2.1 Primary Recovery Stage

The extraction of oil can occur through either natural means or through the utilisation of synthetic lift mechanisms, like hydraulic jacks. The oil present in the stores is found inside the reservoir rocks beneath the Earth's surface, namely in microscopic pores, thin fissures, and gaps. During this phase, the reservoir pressure has a crucial function as it facilitates the efficient upward movement of oil. The magnitude of this pressure is contingent upon the interior power and characteristics of the petroleum products (Zhang et al. 2020). This phenomenon encompasses several mechanisms that contribute to the movement and energy of water from a substantial subsurface water flow, known as an aquifer. These mechanisms include sources of energy, including aquifers, dispersed gas caps and development caused by the interaction between sedimentary rock and stream. Due to its exclusive focus on the oil and its sensitivity or accessibility, this approach exhibits restricted extraction capability (Rivera et al. 2019; Alagorni et al. 2015).

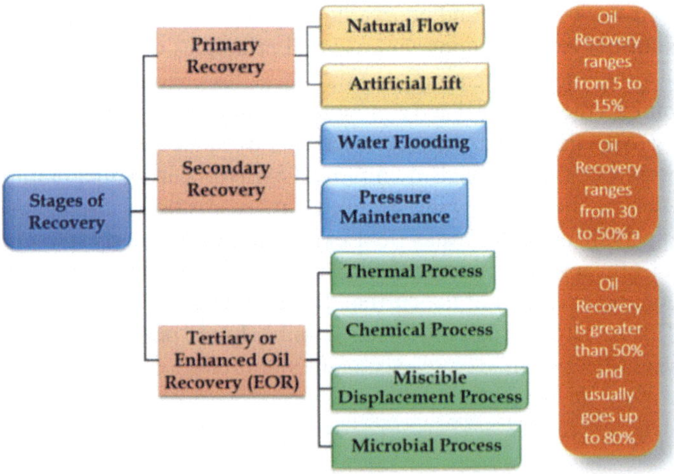

Fig. 12.1 Diagrams illustrating the categorization of the steps involved in oil recovery

The dissolved gas drive mechanism involves the liberation of volatile fluids, resulting in the formation of tiny gas bubbles inside the reservoir. The entrapment of these bubbles within the pore results in their expansion, leading to the displacement of oil. Following this, when the pressure decreases, the gas bubbles coalesce, resulting in a cohesive and swift gas stream that moves up to the top region within the storage area, ultimately resulting in the creation of a gas cap. As the pressure lowers, there is an observed increase in the displacement of the oil, specifically referred to as engine cover (Rivera et al. 2019; Alagorni et al. 2015). Assisted hoisting is commonly employed in cases when the inherent reservoir drive is insufficient to effectively propel the oil up to the top. In this particular scenario, the hydraulic pumps are immersed and employed for the purpose of oil extraction, while the hydraulic lift method involves the injection of the well with pressurised gas in order to reinstate the equivalent pressure (Ganat 2019).

12.2.2 Secondary Recovery Stage

This phase entails the introduction to fill the tank with gas or water with the aim of replenishing the demands or sustaining the flow of oil inside the tank. This process occurs when the reservoir experiences a fall in pressure as a result of the progressive extraction of oil, leading to insufficient pressure to facilitate the upward movement of oil to the surface. At this juncture, the attainment of a crucial threshold is seen, necessitating the use of external energy in the process of oil recovery. Gas injection is mostly employed in conjunction while wearing a gasoline tank in order to move the oil around contained inside reservoirs. The prevention of pore clogging in rocks is achieved by using dissolved gases in the water utilised for the process of flooding. Currently, the viscosity of the oil has been minimised in accordance with the principles of thermodynamics. In this particular scenario, a predetermined pressure gradient is carefully maintained to ensure the optimal displacement of oil, resulting in the maximum quantity being extracted. In succeeding phases, as the ratio of oil to water grows, it becomes economically unfeasible to proceed with the process, therefore necessitating the abandonment of the oil reserve (Alagorni et al. 2015; Anuka et al. 2020; Romero-Zerón 2012). Approximately 50–70% of the global oil that cannot be refined is found in deposits that have depleted using traditional extraction techniques (Oliveira and Sn 2017; Cui et al. 2019; Ali et al. 2019). Therefore, the primary focus of this study is on tertiary extraction technologies, namely those utilised for the extraction of trapped oil from both extant and abandoned deposits.

12.2.3 Tertiary Recovery Stage (Enhanced Oil Recovery)

The primary aim of this particular phase is to optimise the utilisation petroleum products remain unrecoverable in preceding phases of oil extraction. This encompasses leftover lubricant oil. The first term pertains to the stationary residual crude oil present within the pores of the reservoir, whereas the later term pertains to the petroleum oil that remains unremoved owing to the creation of a heterogeneous type. This procedure is employed in industries characterised by limited permeability, heavy oil, or uneven fault lines. Its purpose is to modify either the properties of the reservoir fluid or the features of the rock, in order to make the reservoir more suitable for extraction. The validity of the economic aspects of this strategy must be ensured. Hence, in order to ascertain the most suitable EOR method for a given reservoir, it is imperative to conduct a thorough assessment such as assessment, characterising reservoirs, planning, modelling, and simulating. This stage enables the improvement of crude oil's ability to move, the properties of the oil–water boundary, and the efficiency of both small-scale and large-scale cleaning. Additional comprehensive investigation and experimentation are necessary to thoroughly examine the characteristics of the fluid, reservoir rocks, and specific processes involved in this particular stage (Zhang et al. 2020; Rivera et al. 2019; Muggeridge et al. 2014).

During this phase, the extraction of oil can be achieved through various thermal techniques, including both in situ combustion or fire flooding (Nikolova and Gutierrez 2020; Saravanan et al. 2020) and flooding with steam (Mokheimer et al. 2019; Huang et al. 2018; Saravanan et al. 2020; Pang et al. 2015; Kirmani et al. 2021) are examples of flooding. Additionally, chemical methods involving, surfactants, alkaline displacement and polymers (Saravanan et al. 2020; Gbadamosi et al. 2018; Sheng 2014; Mandal 2015; Kokal and Al-Kaabi 2010; Sun et al. 2020) can be employed. Furthermore, miscible displacement methods utilising substances such as N_2, CO_2, kerosene, flue gases, petrol, benzene, and others (Saravanan et al. 2020; Alamooti and Malekabadi 2018) can be utilised. Lastly, the microbial approach According to a research conducted by Alvarado and Manrique (2010), it was shown that sandstone reservoirs have a higher frequency of utilisation of these technologies compared to other geological types, including turbiditic and carbonate deposits. Figure 12.2 illustrates the distribution of assignments involving EOR across different lithologies. It is evident from the figure that gas injection techniques have been less often utilised in comparison to chemical and thermal procedures for the individual lithologies. Nevertheless, the implementation of these techniques, excluding MEOR, necessitates significant usage of energy for steam production, the utilisation of expensive chemicals, a substantial amount of potable water, and costly machinery to support infrastructure. These factors contribute to the escalation of oil prices and the accompanying natural hazards (Nazina et al. 2020). Contemporary methodologies, example, acoustic stimulations (Rezaei et al. 2018), seismic stimulations (Wooden 2021), as well as electromagnetic techniques (Shafiai and Gohari 2020), are now being employed. The aforementioned approaches have been extensively described in other sources, whereas the present article only concentrates on the microbiological approach.

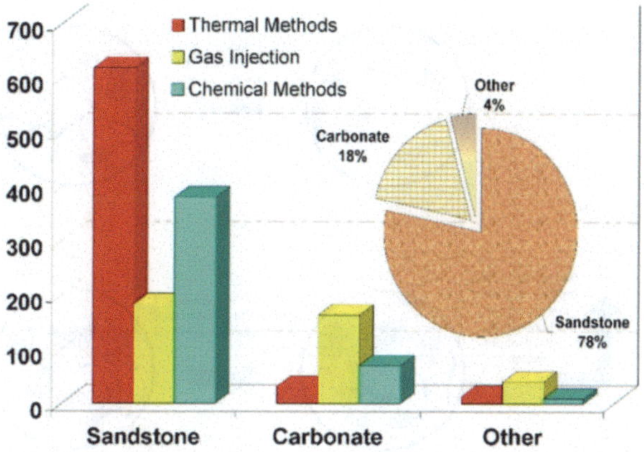

Fig. 12.2 Utilisation of tertiary EOR techniques through lithology. Adapted from Alvarado and Manrique (2010)

12.2.3.1 Microbial Method (MEOR)

An essential resource, MEOR for industries to apply these method in old mature fields. Sadly, not all naturally occurring micobes can guarantee oil recovery. The limited comprehension of the microbial activity process raises doubts about the feasibility of applying microbial expertise to oil deposits via storage tank engineering concepts. While there is a wealth of theoretical information on MEOR, its current low acceptance can be attributed to a number of factors, including a absent of standardised outcomes of the field, a dearth of post-trial investigations, an incomplete knowledge of the process (ecology), and a lack of organised study that offers compelling proof of the procedure (Saikia et al. 2013). MEOR can be put into practise because of its exceptionally low operational expenses (Zhang et al. 2020). The sole microorganisms that are diminutive enough to endure conditions of extreme heat, cold, salinity, and pressure, as well as the ability to produce essential metabolic products, are called extremophiles. Some examples of these microorganisms are halophilic bacteria such as Halobacterium sp. and Salicolamarasensis sp., piezophilic bacteria including Thermococcus barophilus, Shewanella sp., and Photobacterium sp., alkaliphilic bacteria like Natronococcus occultus, thermophilic bacteria such as Pyrolobus fumarii and Thermus aquaticus, and polyextremophile bacteria like Halonatronum saccharophilum and Natranaerobius thermophilus (Parihar and Bagaria 2020; Moreno et al. 2013; Irwin 2020; Varjani and Upasani 2019). The MEOR researchers have shown interest in these naturally occurring bacteria that consume hydrocarbons, are non-virulent, and are well-suited to adapt to the oil environment (Volk and Hendry 2010; Yi et al. 2018; Shibulal et al. 2018).

Figure 12.3 shows a summary of the MEOR procedure. As demonstrated, oil is still present in the reservoir substratum even if traditional procedures are applied later. When bacteria interact with accumulated oil, they produce biochemical byproducts that change the rock/sand grain wettability and Interfacial Tension (IFT), which in turn promotes emulsion formation and improved mobility. Water can travel a

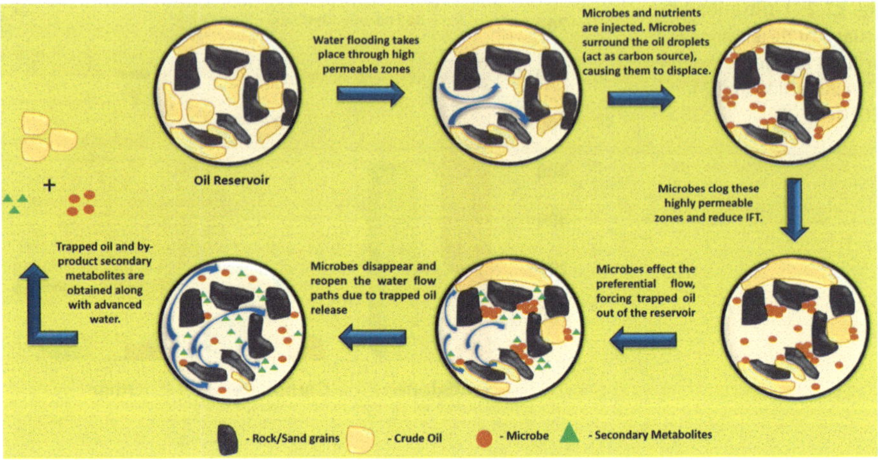

Fig. 12.3 Visual representation of a summary of the MEOR procedure

preferred route to the locations targeted by thieves thanks to the extremely porous zones. The water is then forced to low-permeability places by bacteria that cause "selective plugging," which disrupts the channel's natural flow. This water flows laterally to assist control surplus water production and improve the massive effectiveness of the petroleum by decreasing the remaining oil concentration. During water flooding, the microbial consortia clusters have the ability to vanish and shift, which stimulates the formation of new flow channels in a dynamic manner (Goudarzi et al. 2014; Wood 2019; Kögler et al. 2021).

The objective of this step is to analyse and classify microbial communities, as well as to comprehend their methods of cultivation. The effectiveness of this biological MEOR method relies on various parameters including depth, oil viscosity, pH, permeability, pressure, temperature, crude oil specific gravity, brine salinity, water reduction, porosity, wax content, residual oil saturation, and the microbial species present (Oliveira and Sn 2017; Saravanan et al. 2020; Gaol et al. 2021).

The primary rationales for employing the MEOR technology are oil displacement and enhancement (Shibulal et al. 2014). In the previous method, the grease layers are moved out of the vessels that allow fluid to pass through, resulting in a change in the oil–water interfacial tension (IFT) due to microbial activities. These activities reduce the viscosity of the oil, improve the efficiency of the sweeping process, and restore the pressure in the reservoir (Joshi et al. 2008; Niu et al. 2020). In this method, the reservoir's heavy oil undergoes degradation by microbiological processes, resulting in the conversion of sulphur and heavy metals and the production of lighter oil (Glagoleva and Kapustin 2018). Petroleum and natural gas companies in the modern era consider MEOR as a promising field of study with the opportunity to extract 377 billion barrels of untapped oil (Shibulal et al. 2014; Prabhakar and Melnik 2017).

MEOR, which stands for Microbial Enhanced Oil Recovery, has gained significant attention in the field of research in recent years. It has been anticipated for a

Table 12.1 Biopolymers employed in MEOR process with their micro organism (Adapted from Wu et al. 2022)

Biopolymers	Microbes	Tertiary recovery (%)
Xanthan gum	*Xanthomonas campestris*	42.9
Diutan gum	*Sphingomonas* sp.	20.9
Dextran	*Pseudomonas stutzeri*	13.56
Levan	*Bacillus licheniformis*	45
Pullulan	*Aureobasidium pullulans*	9.41
Scleroglucan	*Sclerotium fungi*	–
Welan gum	*Alcaligenes* sp	34.5
–	*Rhizobium viscosum*	25.7

long time that MEOR may provide a cost-effective method for improving oil extraction processes (Lee et al. 2020). Field studies have demonstrated that the expected decline in oil output may be reversed by the application of MEOR. This is due to the development of microorganisms and the resulting metabolites, which can have diverse effects on the properties of both the reservoir rock and the crude oil (Nikolova and Gutierrez 2020). Research on the extraction of MEOR from unusual oil and gas resources, such as tar sands, oil sands, and bituminous sands, is now underway. These sources provide unique ecological challenges, such as water shortages and high concentrations of hydrocarbons (Miller and Sorrell 2014). The impact of microbial by-products is similar to that of substances that were used before or those being proposed to be used in EOR, with the exception of biomass. Thus, in essence, MEOR is not a groundbreaking and original method for recovering oil. It is worth noting that the main distinction between MEOR and EOR lies in the introduction of chemicals into the reservoirs. In order to achieve a higher level of efficiency compared to EOR, MEOR is often conducted on-site to save logistical expenses (Nikolova and Gutierrez 2020; Song et al. 2016).Table 12.1 depicts the biopolymers employed in MEOR process along with their producing microbes and additional oil recovery percentages.

12.3 Fungi in EOR

Microbial improved oil recovery significantly contributes to the retrieval of heavy oils. However, the focus has primarily been on using bacteria, with less emphasis on exploring the possibilities of fungus. This study examined the effectiveness of fungal extracellular enzymes in converting heavy oil fractions into lighter molecules by biotransformation and biodegradation. Additionally, we assessed the practicality of using these enzyme preparations to improve oil recovery.

Although the utilisation of fungi in enhanced oil recovery has been limited, certain filamentous fungi such as Scedosporium sp. are widely distributed saprotrophs that are frequently found in soil, sewage, waste materials, and water bodies contaminated by environmental pollutants (Cooke and Kabler 1955). These fungi have demonstrated effective degradation of crude oil. They generate extracellular enzymes that have limited substrate specificity, leading to the breakdown of diverse substances, primarily of organic nature, into less harmful forms. Filamentous fungi offer distinct advantages for the restoration and remediation of contaminated environments. Firstly, these fungi can penetrate soil aggregates that are internally porous, anoxic, and hydrophobic, thereby effectively addressing hydrocarbon contamination. Secondly, most fungi are capable of tolerating both dry and osmotic conditions, enabling them to thrive across a wide range of temperatures. This adaptability has been documented in various studies (Smith et al. 2010; Mikolasch et al. 2019; Martínez-Gallardo et al. 2020).Thirdly, filamentous fungi possess the ability to convert complex aliphatic and aromatic compounds found in crude oil into less toxic end-products. Lastly, these fungi remain active throughout the entire degradation process, ensuring sustained effectiveness (April et al. 1999).

Fungi and other microorganisms play a crucial role in oil production, particularly in extracting residual oil that is trapped. These microorganisms generate various metabolites such as biosurfactants, biopolymers, acid gases, and solvents, which possess the ability to eliminate both heavy and light crude oil. Regardless of the scenario, the process of oil recovery always relies on microbial metabolites (Yakimov et al. 1997). The biosurfactant-enhanced oil recovery (EOR) technique involves the synthesis of biosurfactants by specialised microorganisms using low-cost raw materials, which are then injected into the oil reservoir (McInerney et al. 2007). Biosurfactants have a primary role in oil recovery by primarily reducing the interfacial tension between oil and water. This leads to a decrease in the capillary effect and an increase in the capillary number (Xu 2005). The second process involves the production of micelles, which serves as a physical mechanism for mobilising oil through the movement of an aqueous phase. Both of these effects arise from the existence of hydrophilic and hydrophobic structural components, which have an attraction to the water phase and the oil phase, respectively (Simpson et al. 2011). The microbial surface active agents surpass conventional ones due to their ability to reduce surface and interfacial tension, while also being biocompatible (Banat et al. 2010). Compared to surfactants that are commercially synthesised, these substances are biodegradable, non-toxic, and environmentally benign. They exhibit minimal irritation (Pornsunthorntawee et al. 2008) and retain their characteristics even under high temperatures (Ron and Rosenberg 2002). Their activities are quite particular as a result of the functional groupings they possess (Whang et al. 2008). They also need a lower amount of initial financial commitment as compared to chemically produced surfactants (Morita et al. 2007). Microbially-produced biosurfactants decrease the interfacial tension between oil and water and reduce the viscosity of oil, facilitating the release of trapped residual oil from reservoirs (Bass and Lappin-Scott 1997; Li et al. 2002). This cost-effective approach leads to a boost in oil output, potentially achieving self-sufficiency and reducing reliance on imported oil, so producing cash.

References

Alagorni AH, Yaacob ZB, Nour AH (2015) An overview of oil production stages: enhanced oil recovery techniques and nitrogen injection. Int J Environ Sci Dev 6:693–701

Alamooti AM, Malekabadi FK (2018) An introduction to enhanced oil recovery. In: Fundamentals of enhanced oil and gas recovery from conventional and unconventional reservoirs. Elsevier, Amsterdam, The Netherlands, pp 1–40

Ali N, Wang F, Xu B, Safdar B, Ullah A, Naveed M, Wang C, Rashid MT (2019) Production and application of biosurfactant produced by bacillus licheniformis Ali5 in enhanced oil recovery and motor oil removal from contaminated sand. Molecules 24:4448

Alvarado V, Manrique E (2010) Enhanced oil recovery: an update review. Energies 3:1529–1575

Anuka A, Falode O, Ogunshe A, Udie C (2020) An overview of microbial enhanced oil recovery in Nigeria. In: Proceedings of the society of petroleum engineers—SPE Nigeria annual international conference and exhibition

April TM, Foght JM, Currah RS (1999) Hydrocarbon-degrading filamentous fungi isolated from flare pit soils in northern and western Canada. Can J Microbiol 46(1):38–49

Banat IM et al (2010) Microbial biosurfactants production, applications and future potential. Appl Microbiol Biot 87:427–444

Bass CJ, Lappin-Scott HM (1997) The bad guys and the good guys in petroleum microbiology. Oilfield Rev 9:17–25

Bi Y, Xiu J, Ma T (2019) Application potential analysis of enhanced oil recovery by biopolymer-producing bacteria and biosurfactant-producing bacteria compound flooding. Appl Sci 9:5119

Cooke WB, Kabler P (1955) Isolation of potentially pathogenic fungi from polluted water and sewage. Public Health Rep 70(7):689

Cui K, Zhang Z, Zhang Z, Sun S, Li H, Fu P (2019) Stimulation of indigenous microbes by optimizing the water cut in low permeability reservoirs for green and enhanced oil recovery. Sci Rep 9:1–12

Ganat T (2019) Pumping system of heavy oil production. Process Heavy Crude Oils Chall Oppor 12

Gaol CL, Ganzer L, Mukherjee S, Alkan H (2021) Parameters govern microbial enhanced oil recovery (MEOR) performance in real-structure micromodels. J Pet Sci Eng 205:108814

Gbadamosi AO, Kiwalabye J, Junin R, Augustine A (2018) A review of gas enhanced oil recovery schemes used in the North Sea. J Pet Explor Prod Technol 8:1373–1387

Glagoleva OF, Kapustin VM (2018) Physicochemical aspects of primary oil processing technology (Review). Pet Chem 58:1–7

Goudarzi A, Almohsin A, Varavei A, Delshad M, Bai B, Sepehrnoori K (2014) New experiments and models for conformance control microgels. In: Proceedings of the conference improved oil recovery symposium, Tulsa, OK, USA, pp 1–17

Haq B (1946) The role of microbial products in green enhanced oil recovery: acetone and butanone. Polymer 2021:13

Huang S, Cao M, Cheng L (2018) Experimental study on the mechanism of enhanced oil recovery by multi-thermal fluid in offshore heavy oil. Int J Heat Mass Transf 122:1074–1084

Irwin JA (2020) Overview of extremophiles and their food and medical applications. Physiol Biotechnol Asp Extrem 65–87

Joshi S, Bharucha C, Jha S, Yadav S, Nerurkar A, Desai AJ (2008) Biosurfactant production using molasses and whey under thermophilic conditions. Bioresour Technol 99:195–199

Kirmani FUD, Raza A, Gholami R, Haidar MZ, Fareed CS (2021) Analyzing the effect of steam quality and injection temperature on the performance of steam flooding. Energy Geosci 2:83–86

Kögler F, Dopffel N, Mahler E, Hartmann FSF, Schulze-Makuch D, Visser F, Frommherz B, Herold A, Alkan H (2021) Influence of surface mineralogy on the activity of Halanaerobium sp. during microbial enhanced oil recovery (MEOR). Fuel 290:119973

Kokal S, Al-Kaabi A (2010) Enhanced oil recovery: challenges & opportunities. World Pet Counc Off Publ 64:64–69

Lee KS, Kwon TH, Park T, Jeong MS (2020) Introduction. In: Theory and practice in microbial enhanced oil recovery. Elsevier, Amsterdam, The Netherlands, pp 1–25

Li Q, Kanga C, Wang H, Liu C, Zhang C (2002) Application of microbial enhanced oil recovery technique to Daqing Oilfield. Biochem Eng J 11:197–199

Mandal A (2015) Chemical flood enhanced oil recovery: a review. Int J Oil Gas Coal Technol 9:241–264

Martínez-Gallardo MR, López MJ, Jurado MM, Suárez-Estrella F, López-González JA, Sáez JA, Moral R, Moreno J (2020) Bioremediation of Olive Mill Wastewater sediments in evaporation ponds through in situ composting assisted by bioaugmentation. Sci Total Environ 703:135537

McInerney MJ, Voordouw G, Jenneman GE, Sublette KL (2007) Oil field microbiology. In: Hurst CJ, Crawford RL, Garland JL, Lipson DA, Mills AL, Stetzenbach LD (eds) Manual of environmental microbiology. ASM, Washington, DC, pp 898–911

Mikolasch A, Donath M, Reinhard A, Herzer C, Zayadan B, Urich T, Schauer F (2019) Diversity and degradative capabilities of bacteria and fungi isolated from oil-contaminated and hydrocarbon-polluted soils in Kazakhstan. Appl Microbiol Biotechnol 103:7261–7274

Miller RG, Sorrell SR (2014) The future of oil supply. Philos Trans R Soc A Math Phys Eng Sci 372

Mokheimer EMA, Hamdy M, Abubakar Z, Shakeel MR, Habib MA, Mahmoud MA (2019) comprehensive review of thermal enhanced oil recovery: techniques evaluation. J Energy Resour Technol Trans ASME 141

Moreno MDL, Pérez D, García MT, Mellado E (2013) Halophilic bacteria as a source of novel hydrolytic enzymes. Life 3:38–51

Morita T, Konishi M, Fukuoka T, Imura T, Kitamoto D (2007) Physiological differences in the formation of the glycolipid biosurfactants, mannosylerythritol lipids, between Pseudozymaantarctica and Pseudozymaaphidis. Appl Microbiol Biot 74:307–315

Muggeridge A, Cockin A, Webb K, Frampton H, Collins I, Moulds T, Salino P (2014) Recovery rates, enhanced oil recovery and technological limits. Philos Trans R Soc A Math Phys Eng Sci 372

Nazina T, Sokolova D, Grouzdev D, Semenova E, Babich T, Bidzhieva S, Serdukov D, Volkov D, Bugaev K, Ershov A et al (2020) The potential application of microorganisms for sustainable petroleum recovery from heavy oil reservoirs. Sustainability 12:15

Nikolova C, Gutierrez T (2020) Use of microorganisms in the recovery of oil from recalcitrant oil reservoirs: current state of knowledge, technological advances and future perspectives. Front Microbiol 10:2996

Niu J, Liu Q, Lv J, Peng B (2020) Review on microbial enhanced oil recovery: mechanisms, modeling and field trials. J Pet Sci Eng 192:107350

Oliveira X, Sn S (2017) The challenge of the use the microorganisms in enhance oil recover. Recent Adv Petrochem Sci 1:0034–0036

Pang Z, Liu H, Zhu L (2015) A laboratory study of enhancing heavy oil recovery with steam flooding by adding nitrogen foams. J Pet Sci Eng 128:184–193

Parihar J, Bagaria A (2020) The extremes of life and extremozymes. Divers Perspect 3:2581–3226

Pornsunthorntawee O, Wongpanit P, Chavadej S, Abe M, Rujiravanit R (2008) Structural and physicochemical characterization of crude biosurfactant produced by Pseudomonas aeruginosa SP4 isolated from petroleum-contaminated soil. Biores Technol 99:1589–1595

Prabhakar S, Melnik R (2017) Wettability alteration of calcite oil wells: influence of smart water ions. Sci Rep 7:1–9

Rezaei Dehshibi R, Mohebbi A, Riazi M, Niakousari M (2018) Experimental investigation on the effect of ultrasonic waves on reducing asphaltene deposition and improving oil recovery under temperature control. Ultrason Sonochem 45:204–212

Rivera MAH, Vasconcellos JM, Morales MEO (2019) Factors affecting microbial enhanced oil recovery (MEOR). In: Proceedings of the 25th Pan-American conference of naval engineering—COPINAVAL. Springer, Berlin/Heidelberg, Germany, pp 375–384

Romero-Zerón L (2012) Introduction to enhanced oil recovery (EOR) processes and bioremediation of oil-contaminated sites. IntechOpen, London, UK

Romero-Zeron L (2016) Chemical enhanced oil recovery (cEOR)—a practical overview. IntechOpen, London, UK

Ron EZ, Rosenberg E (2002) Biosurfactants and oil bioremediation. Curr Opin Biotechnol 13:249–252

Safdel M, Anbaz MA, Daryasafar A, Jamialahmadi M (2017) Microbial enhanced oil recovery, a critical review on worldwide implemented field trials in different countries. Renew Sustain Energy Rev 74:159–172

Saikia U, Bharanidharan R, Vendhan E, Kumar Yadav S, Siva Shankar S (2013) A brief review on the science, mechanism and environmental constraints of microbial enhanced oil recovery (MEOR). Int J Chem Techol Res 5:1205–1212

Saravanan A, Kumar PS, Vardhan KH, Jeevanantham S, Karishma SB, Yaashikaa PR, Vellaichamy P (2020) A review on systematic approach for microbial enhanced oil recovery technologies: opportunities and challenges. J Clean Prod 258:120777

Shafiai SH, Gohari A (2020) Conventional and electrical EOR review: the development trend of ultrasonic application in EOR. J Pet Explor Prod Technol 10:2923–2945

She H, Kong D, Li Y, Hu Z, Guo H (2019) Recent advance of microbial enhanced oil recovery (MEOR) in China. Geofluids 2019:1–16

Sheng JJ (2014) A comprehensive review of alkaline-surfactant-polymer (ASP) flooding. Asia Pac J Chem Eng 9:471–489

Shibulal B, Al-Bahry SN, Al-Wahaibi YM, Elshafie AE, Al-Bemani AS, Joshi SJ (2014) Microbial enhanced heavy oil recovery by the aid of inhabitant spore-forming bacteria: an insight review. Sci World J 2014:1–12

Shibulal B, Al-Bahry SN, Al-Wahaibi YM, Elshafie AE, Al-Bemani AS, Joshi SJ (2018) Microbial-Enhanced Heavy Oil Recovery under Laboratory Conditions by Bacillus firmus BG4 and Bacillus halodurans BG5 Isolated from Heavy Oil Fields. Colloids Interfaces 2:1

Simpson D, Natraj N, McInerney M, Duncan K (2011) Biosurfactant-producing Bacillus are present in produced brines from Oklahoma oil reservoirs with a wide range of salinities. Appl Microbiol Biot 91:1083–1093

Smith IM, Cook DR, Smith BP (2010) Water mites (Hydrachnidiae) and other arachnids. In: Ecology and classification of North American freshwater invertebrates. Academic Press, pp 485–586

Song W, Ma D, Wu K, Wei X, Xu Y (2016) A review study: development of microbial enhancing oil recovery technology. In: Proceedings of the 2016 5th international conference on environment, materials, chemistry and power electronics, Zhengzhou, China, pp 619–625

Sun C, Guo H, Guo H, Guo H, Li Y, Jiang G, Ma R (2020) Alkali effect on alkali-surfactant-polymer (ASP) flooding enhanced oil recovery performance: two large-scale field tests' evidence. J Chem 2020

Varjani S, Upasani VN (2019) Evaluation of rhamnolipid production by a halotolerant novel strain of Pseudomonas aeruginosa. Bioresour Technol 288:121577

Volk H, Hendry P (2010) 3° oil recovery: fundamental approaches and principles of microbially enhanced oil recovery. In: Handbook of hydrocarbon and lipid microbiology. Springer, Berlin/Heidelberg, Germany, pp 2727–2738

Whang LM, Liu PWG, Ma CC, Cheng SS (2008) Application of biosurfactants, rhamnolipid, and surfactin, for enhanced biodegradation of diesel-contaminated water and soil. J Hazard Mater 151:155–163

Wood DA (2019) Microbial improved and enhanced oil recovery (Mieor): review of a set of technologies diversifying their applications. Adv Geo Energy Res 3:122–140

Wooden B (2021) Seismic stimulation: an eco-friendly, effective EOR alternative. Available online: https://jpt.spe.org/seismic-stimulation-eco-friendly-effective-eor-alternative. Accessed on 19 June 2021

Wu B, Xiu J, Yu L, Huang L, Yi L, Ma Y (2022) Research advances of microbial enhanced oil recovery. Heliyon

Xu W (2005) Experimental investigation of dynamic interfacial interactions reservoir conditions. MS thesis, Louisiana State Univ., Baton Rouge, LA 70803, USA
Yakimov MM, Amro MM, Bock M, Boseker K, Fredrickson HL, Kessel DG, Timmis KN (1997) The potential of Bacillus licheniformis strains for in situ enhanced oil recovery. J Pet Sci Eng 18:147–160
Yi LN, Li ZP, Liu L, Bi YQ, Wang XT, Yi JP (2018) Functional microbial stimulation for oil recovery enhancement based on microbial community analysis. Biotechnol Biotechnol Equip 32:1468–1476
Zhang J, Gao H, Xue Q (2020) Potential applications of microbial enhanced oil recovery to heavy oil. Crit Rev Biotechnol 40:459–474

Chapter 13
Production Impact Due Deposition of Hydrocarbons in Oil and Gas Industries

Amit Kumar, Prathibha Pillai, Md. Merajul Islam, and Neha Saxena

Abstract Oil and gas industry deals with the supply of crude oil and gas for the fulfillment of energy requirement of the world. The crude oil and gas are recovered from subsurface reservoirs and supplied to refineries and petrochemical industries. The unrestricted flow of crude oil and gases through the pores of the reservoir, through the flowlines connecting subsurface reservoir and surface facilities, and through pipelines transporting the crude oil to the refineries is important to maintain the energy needs of the world. Crude oil, which is a mixture of hydrocarbons, can have varying molecular weight ranging from 100 to 1000, and have different flowability in reservoir pores and flowlines. The heavier crude oil will have most difficulty in flowing due to presence of higher amount of asphaltene, paraffin, and resins. Heavy crude oil also suffers from the deposition of these hydrocarbon components, which restricts the available path for flow and hampers the production from the reservoir, and transport of the crude oil through pipelines. The deposition also causes a significant increase in the required pressure for maintaining the flow through the pipelines. The extended deposition of hydrocarbons can also completely block a flow path and halt the production of crude oil. The hampered production due to deposition issues may lead to missed opportunities to capitalize on favorable market conditions or to meet production targets. The overall profitability of oil and gas operations can be impacted as a result. The chapter deals with various strategies, including chemical inhibitors, mechanical cleaning methods, pigging operations (pipeline cleaning tools), employed by the oil and gas industries to mitigate these production impacts.

A. Kumar (✉)
Department of Petroleum Engineering and Geoengineering, Rajiv Gandhi Institute of Petroleum Technology, Jais, Uttar Pradesh, India
e-mail: amit_02_03@yahoo.in

P. Pillai
Halliburton, Mumbai, India

Md. M. Islam · N. Saxena
School of Basic Sciences and Technology, IIMT University, Meerut, India

S. Banerjee et al. (eds.), *Environmental Friendly Green Technologies for Improvement of Heavy Crude Oil Flow Assurance*, Environmental Science and Engineering,
https://doi.org/10.1007/978-3-031-86701-9_13

Keywords Hydrocarbon deposition · Production · Heavy crude oil · Plugging · Mitigation

13.1 Introduction

The oil and gas industry is one the most important industrial sectors, as it covers a significant amount of energy requirement for development of the world. The oil and gas industry is a dynamic and continuously evolving sector that consists of operations, such as explorations, drilling, production, and refining. The industry is also responsible for also distribution of energy for fulfilling the needs of every facet of human life. However, within this multifaceted landscape, there exists a significant challenge, i.e., the deposition of hydrocarbons.

Hydrocarbon deposition is the accumulation of a few components present in the crude oil, such as paraffin, and asphaltene in the different parts of the production and transportation process. The issue of hydrocarbon deposits is important for the industry as it can have far-reaching consequences for the main goal of the industry. In the following sections, the multifaced impact of the deposition of hydrocarbon has been discussed and the types of hydrocarbon deposits, and their detrimental effects on flow rates, equipment efficiency, and system integrity are described. The economic implications, maintenance challenges, and the necessity for proactive mitigation strategies will also be discussed.

Hydrocarbon deposition is a major problem for the oil and gas industry, and it is important to understand and manage it. Hydrocarbon deposition can reduce production yields, increase operational costs, and harm the environment. The oil and gas industry is continuously working to improve its production processes and ensure sustainable resource extraction by addressing this challenge.

This chapter will provide a comprehensive overview of hydrocarbon deposition in the oil and gas industry, its complexities, and the importance of proactive management. The role of emerging technologies in mitigating their impacts have been discussed. By understanding the challenges posed by hydrocarbon deposition, a more efficient, environmentally responsible, and sustainable energy future can be reached.

13.2 Types of Hydrocarbon Deposits

The crude oil present in the reservoirs across the globe is a mixture of hydrocarbon and non-hydrocarbon components. The compositions of crude oil in different reservoirs can have significant variations, as well as, such variations can also exist within a single reservoir (Speight 2015). The compositional variation of different crude oils occurs because of the variation in the depositional environment of the formation, the geological maturity of the reservoir, and the migration path of the crude oil from the

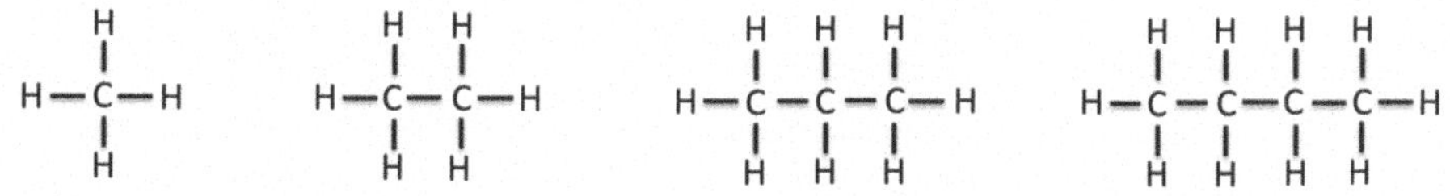

Fig. 13.1 Paraffinic components of crude oil

source rock to the reservoir rock. It is impossible to identify and measure each of the thousands of components that can be found in a reservoir of crude oil. Thus, the variation in the composition of the crude oils can be identified based on the variation of the different classes of hydrocarbon present in the crude oil. These classifications of the hydrocarbons are based on their structure. Major classes of hydrocarbons are paraffins, olefins, naphthenes, and aromatics.

Paraffin are saturated hydrocarbons, also known as alkanes, with the general molecular formula of C_nH_{2n+2}. These hydrocarbons contain only single covalent bonds between the carbon atoms and the hydrogen atoms, thus are called saturated hydrocarbons. Due to their saturated structure, they are also less reactive than other components present in crude oil, and thus also contribute to the highest composition of the crude oil. Figure 13.1 shows the paraffinic components of crude oil.

The paraffinic components of higher molecular weight are separately categorized as paraffin wax and they are a complex mixture, consisting of straight, branched, and cyclic chains, that can be either solid or liquid at room temperature (Yang et al. 2013). They can be divided in to two groups: macrocrystalline and microcrystalline waxes (Mohamed et al. 2017). Macrocrystalline waxes have straight carbon chain lengths ranging from C16 to C40 (n-alkanes) and crystallize in a platelet-like or a needle-like structure. Microcrystalline waxes have a high percentage of iso-paraffin hydrocarbons (hydrocarbons with branched carbon chains) and naphthenic rings (cyclic hydrocarbons with a saturated ring structure) with carbon chain lengths ranging from C30 to C60 and crystallize in an amorphous-like structure (without a definite shape or form) (Speight 2014).

Olefins are unsaturated hydrocarbons, also known as alkenes, with the general molecular formula of C_nH_{2n}. They are characterized by the presence of at least one carbon–carbon double bond (C=C) in their chemical structure. This double bond introduces a degree of unsaturation, meaning that olefins are not "saturated" with the maximum number of hydrogen atoms like alkanes (paraffins). This double bond is responsible for their reactivity and distinctive chemical properties. This unsaturation leads to increased reactivity, allowing olefins to participate in various chemical reactions, including addition reactions. Olefins also exhibit geometric isomerism, also known as cis–trans isomerism, due to the restricted rotation around the carbon–carbon double bond. This can lead to different spatial arrangements of atoms, resulting in different isomers with distinct chemical properties. Olefins are essential components of the petrochemical industry. They are typically produced from crude oil and natural gas through processes such as steam cracking and catalytic cracking. Olefins are crucial building blocks for a wide range of products and materials, making them

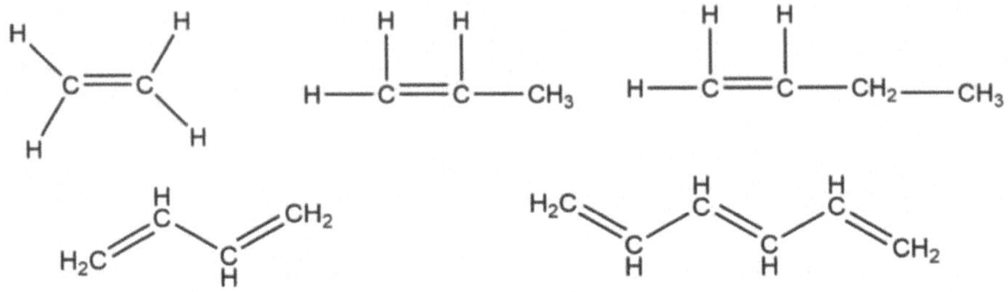

Fig. 13.2 Olefinic components of crude oil

a fundamental part of the chemical and manufacturing industries. Figure 13.2 shows the olefinic components of crude oil.

Naphthanes, also known as cycloalkanes or naphthenes, are characterized by having carbon atoms arranged in a closed ring structure with single covalent bonds. Naphthanes are considered saturated hydrocarbons because they do not contain any carbon–carbon double or triple bonds, unlike olefins or alkynes. Naphthanes have a ring structure, where carbon atoms are bonded to each other to form a closed, cyclic arrangement. The simplest naphthane, cyclohexane, consists of a six-carbon ring. Naphthanes are also saturated hydrocarbons because they contain only single bonds between carbon atoms and are fully "saturated" with hydrogen atoms. Naphthanes generally have similar physical properties to alkanes. They are typically colorless and odorless, and their physical state (solid, liquid, or gas) at room temperature depends on the number of carbon atoms. Figure 13.3 shows the naphthenic components of crude oil.

Aromatics are organic compounds that contain one or more aromatic rings. An aromatic ring is a planar ring of six carbon atoms, with each carbon atom bonded to one hydrogen atom and one other carbon atom. The carbon atoms in an aromatic ring are also bonded to each other by a delocalized pi-electron cloud, which is what gives aromatics their unique properties. Aromatics are typically colorless or pale yellow in color and have a strong, characteristic odor. They are also insoluble in water, but soluble in organic solvents. Figure 13.4 shows the aromatic components of crude oil.

The higher molecular weight napthanes and aromatics and their derivatives with very complex structures can be categorized as asphaltenes. They are the organic compounds that are found in crude oil and are the heaviest and most polar fraction of

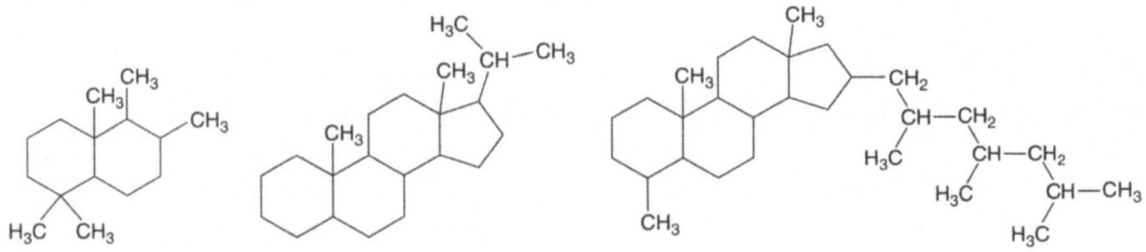

Fig. 13.3 Naphthenic components of crude oil

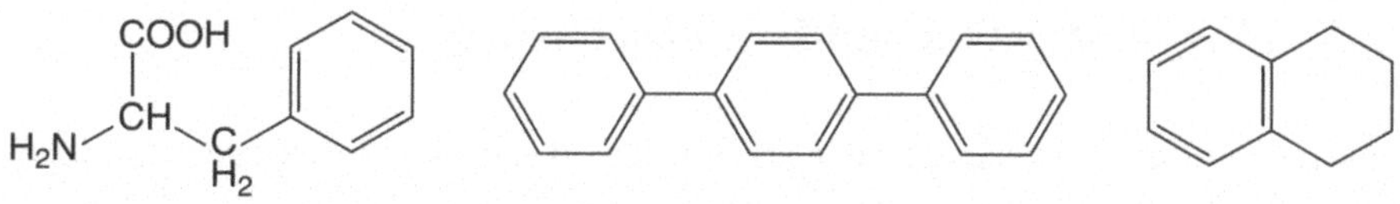

Fig. 13.4 Aromatic components of crude oil

crude oil, and are insoluble in light alkanes such as pentane and heptane, but soluble in aromatic solvents such as toluene and benzene. Asphaltenes are typically black or dark brown in color and have a high carbon content. The exact chemical composition of asphaltenes is not fully understood, but they are thought to be composed of polyaromatic carbon ring units with oxygen, nitrogen, and sulfur heteroatoms, combined with trace amounts of heavy metals, particularly chelated vanadium and nickel, and aliphatic side chains of various lengths (Zuo et al. 2019). Figure 13.5 shows the molecular structure of an asphaltene.

Waxy crude oil containing a significant amount of paraffin wax and asphaltenic crude oil containing a significant amount of asphaltenes are the two major categories of crude oil that have issues of hydrocarbon deposition (Thanh et al. 1999). The deposition of wax occurs below wax appearance temperature (WAT) and subsequent deposition of wax can impede the flowability of waxy crude oil and cause production shutdown (Alnaimat and Ziauddin 2020). Similarly, asphaltene precipitation can block pipelines and production equipment, leading to reduced production and increased operational costs. Avoiding the many issues related to pipeline obstruction and clogging requires a detailed grasp of wax deposition and asphaltene precipitation.

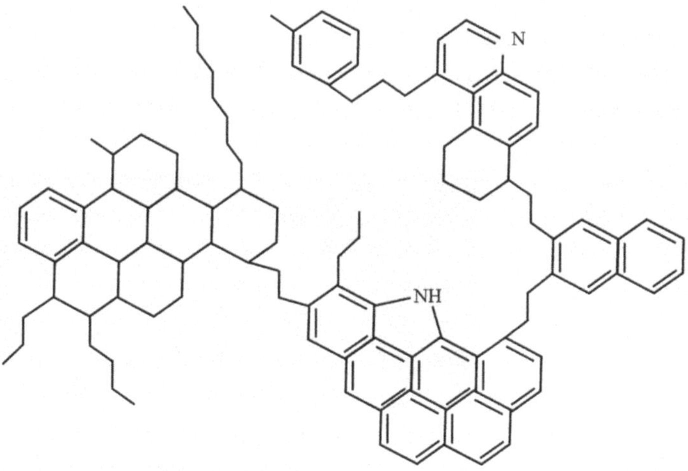

Fig. 13.5 Molecular structure of an asphaltene (Fakher et al. 2020)

13.3 Flow Restriction and Its Consequences

Flow restriction is a critical issue that arises in the oil and gas industry due to the deposition of hydrocarbons such as wax, and asphaltene. Figure 13.6 shows the deposition of wax in a subsea pipeline. As these substances accumulate in pipelines, wellbores, and production equipment, they create significant hindrances to the smooth flow of oil and gas (Sousa et al. 2019). Flow restriction has far-reaching consequences on oil and gas production. Flow restriction leads to a substantial reduction in the flow rates of oil and gas within the production system. This decrease in flow rates can be attributed to the reduction in the cross-sectional area. As hydrocarbon deposits accumulate on the inner surfaces of pipelines, the effective cross-sectional area available for fluid flow is diminished. This results in a decrease in the volume of oil and gas that can pass through the system per unit of time. The decrease in flow rate is also caused due to the increase in frictional resistance. The presence of deposits creates roughness on the pipeline walls, increasing frictional resistance to fluid flow. This increased resistance requires higher pressures to maintain the same flow rates, which can be energy-intensive and lead to equipment strain. Flow restriction can alter the fluid dynamics within the pipelines (Hansen 2023). This can lead to flow instabilities, such as turbulence, which further reduces the flow rates and can potentially damage equipment.

The reduced flow rates resulting from flow restriction have a direct impact on the pressure within the system. Pressure drop, i.e., the decrease in pressure as fluids flow through a pipeline, occurs as a consequence of flow restriction. This pressure drop has significant implications for oil and gas production, as to overcome the elevated frictional resistance and to maintain the desired pressure levels, more energy is required, often in the form of increased pumping or compression. This added energy consumption translates into higher operational costs and reduced overall efficiency. The need for higher pressures and increased energy input can lead to overloading of pumping or compression equipment. This, in turn, can accelerate equipment wear and tear

Fig. 13.6 Wax deposition in a subsea pipeline (Banki et al. 2008)

and necessitate more frequent maintenance and replacements. High-pressure differentials can result in operational challenges and potential safety concerns. Managing and controlling these pressure differentials requires sophisticated monitoring and control systems to prevent catastrophic failures.

Flow restriction due to hydrocarbon deposition also poses risks to the integrity of wellbores and pipelines. The increased pressure differentials resulting from flow restriction can subject the wellbore and pipeline materials to higher stress levels. Over time, this can cause material fatigue and structural damage, potentially leading to leaks or ruptures. Flow restriction can create areas of localized turbulence and altered flow patterns, which may lead to localized corrosion. This corrosion can further compromise the integrity of the equipment and increase maintenance requirements. The cumulative effect of flow restriction on wellbores and pipelines can reduce their overall lifespan. This necessitates more frequent replacements or repairs, adding to the operational costs.

Thus, flow restriction caused by hydrocarbon deposition is a critical challenge in the oil and gas industry. It leads to reduced flow rates, increased pressure drops, and risks to the integrity of production equipment. These consequences necessitate proactive measures to manage and mitigate flow restriction, including regular monitoring, maintenance, and the use of innovative technologies to prevent or remove deposits. The deposition of wax and asphaltene reduces the effective surface area for fluid flow leading to an extreme scenario of complete blockage of the flowline and complete shutdown of production and financial loss to the operating company.

13.4 Efficiency Reduction in Production Equipment

The impact of hydrocarbon deposition on oil and gas production extends beyond flow restriction and pressure drop. It also exerts a significant influence on the efficiency of production equipment, including heat exchangers, separators, and related components. In this section, we delve into how the deposition of hydrocarbons can compromise the efficiency of these essential pieces of equipment.

Heat exchangers play a crucial role in the oil and gas industry, primarily in processes involving temperature control, such as cooling or heating crude oil or gas. Hydrocarbon deposits on the heat exchange surfaces act as insulators, effectively reducing the rate at which heat can be transferred between fluids (Oon et al. 2020). This manifests by reducing the heat transfer efficiency. Deposits create a thermal barrier, leading to reduced heat transfer efficiency. As a result, more energy may be required to achieve the desired temperature changes, increasing operational costs. Slower heat transfer can extend the time needed to process fluids, causing bottlenecks in the production process. This inefficiency can be particularly detrimental in facilities where time-sensitive processes are critical. Reduced heat transfer efficiency can also lead to equipment fouling and corrosion, necessitating more frequent maintenance and cleaning to maintain optimal performance.

Separation equipment, such as separators and coalescers, is essential for separating oil, gas, and water. The presence of hydrocarbon deposits can disrupt these processes by the formation of emulsions and reducing the separation efficiency (Raya et al. 2020). Deposits can lead to emulsion formation, where oil, gas, and water are mixed in undesirable ways. This makes it challenging to separate these components efficiently. The increased viscosity of the fluids caused by deposits can hinder the separation process, resulting in incomplete separation, reduced product quality, and potential equipment fouling. Dealing with emulsions and poor separation efficiency requires additional equipment and energy, further adding to operational costs and reducing overall efficiency.

The inefficiencies caused by hydrocarbon deposition translate into increased energy consumption and operational costs. To compensate for reduced heat transfer efficiency and the challenges in separation equipment, operators may have to resort to increasing the energy input. More energy may be needed to achieve the desired production rates and quality standards. This could involve using additional heating or compression to overcome the hindered processes. Maintenance and cleaning of fouled equipment will also be more frequent and labor-intensive, adding to operational expenses. Frequent fouling and corrosion can lead to equipment deterioration and a shortened operational lifespan, necessitating equipment replacements and further capital expenditure.

Thus, the deposition of hydrocarbons in the oil and gas industry adversely affects the efficiency of production equipment, resulting in increased operational costs, energy consumption, and potential equipment degradation. Effective management and prevention strategies are crucial to mitigating these effects and ensuring the continued performance of production equipment.

13.5 Plugging and Blockages

One of the most disruptive consequences of hydrocarbon deposition in the oil and gas industry is the formation of plugging and blockages within pipelines, wellbores, and production equipment. These blockages can bring production to a halt and often require costly and time-consuming remedial actions. Blockages typically occur due to the buildup of solid or semi-solid hydrocarbon deposits within the production system.

The three stages of phase transformation of waxy crude oil are the aggregation of wax particles in the oil clusters as temperature and pressure change (Chala et al. 2018), the gelation of wax layer, which primarily influences the poor flowability of waxy crude oils (Li et al. 2018), and layer-wise deposition of gel-like wax, which results in flowline blockage that ultimately leads to the complete shutdown of production processes. The continuous deposition of precipitated wax on the surface of the pipeline causes the wax to gel. Figure 13.7 shows the deposition of wax due to reduction in temperature.

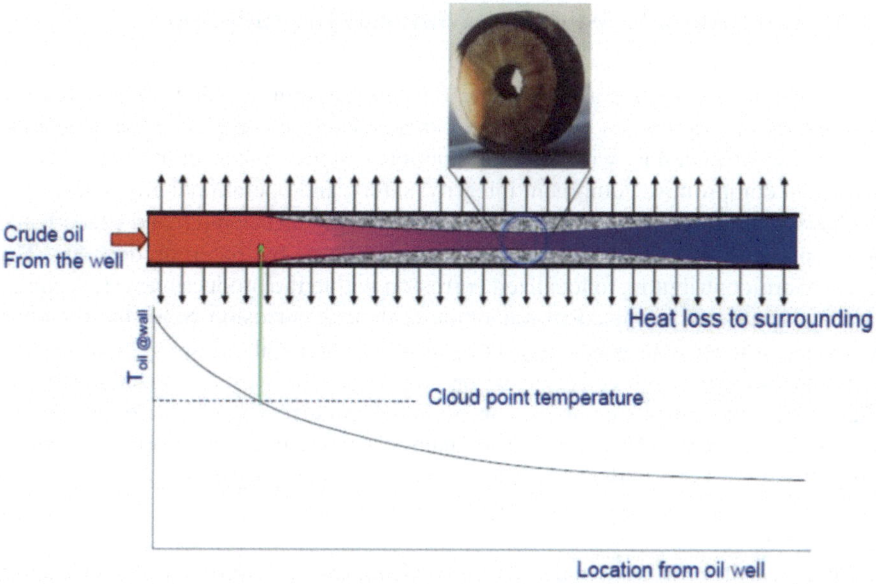

Fig. 13.7 Deposition of wax due to reduction in temperature (Theyab and Diaz 2016)

The process of solid layer development, growth, and deposition on a pipeline's inner surface is known as the wax deposition mechanism. Wax deposition is only possible when the pipe's inner surface temperature is lower than the temperature at which wax appears (WAT). A number of explanations have been put out to explain how wax deposits on pipeline walls occur. Molecular diffusion, shear dispersion, Brownian diffusion, gravity settling, and Soret diffusion are a few of these processes. (Burger et al. 1981; Singhai et al. 1991; Han et al. 2010).

The presence of blockages due to hydrocarbon deposition has serious and immediate consequences for oil and gas production as severe blockages can bring production to a complete standstill, resulting in significant revenue losses. This often necessitates emergency interventions to restore production. The pressure differentials caused by blockages can damage production equipment and infrastructure, leading to costly repairs and replacements. Sudden production halts and equipment damage can pose safety risks to personnel, particularly when dealing with high-pressure systems. Production shutdowns can also result in the release of gases and chemicals into the environment, leading to environmental damage and regulatory compliance issues.

Thus, plugging and blockages resulting from hydrocarbon deposition are among the most disruptive challenges in the oil and gas industry. Preventive measures, vigilant monitoring, and timely intervention are crucial in minimizing their occurrence and mitigating their severe consequences.

13.6 Corrosion, Erosion, and Material Degradation

The deposition of hydrocarbons in the oil and gas industry doesn't just impede flow and reduce equipment efficiency; it can also lead to corrosion, erosion, and the degradation of materials within the production system (Askari et al. 2021). These issues can compromise equipment integrity, safety, and operational efficiency.

Corrosion is a natural chemical process that can be accelerated and exacerbated by the presence of certain substances, including some hydrocarbon deposits. The mechanisms contributing to localized corrosion within the production system can be chemical reactions, pitting corrosion, or under-deposit corrosion. Some hydrocarbon deposits can react with water, oxygen, and other chemicals in the system, leading to the breakdown of protective layers on equipment surfaces. Localized corrosion, often in the form of pits or holes, can occur when corrosive agents attack specific areas of equipment. These pits can weaken structural integrity. Corrosion can be especially aggressive under deposits, where conditions may be ideal for chemical reactions to occur. This type of corrosion is often hard to detect until serious damage has occurred.

Erosion is the process of material removal from surfaces due to the abrasive action of particles suspended in a fluid. When hydrocarbon deposits accumulate within the production system, they can contribute to erosion. Some solid hydrocarbon deposits can act as abrasive particles, causing wear and erosion of equipment surfaces as fluids flow past them. Flow restriction caused by deposits can lead to increased fluid velocities in localized areas, intensifying the erosive forces acting on equipment surfaces.

Given the potential for corrosion, erosion, and material degradation in the presence of hydrocarbon deposits, the choice of materials for equipment and the implementation of corrosion prevention strategies become paramount. Choosing materials that are resistant to corrosion and erosion is critical. Materials like corrosion-resistant alloys are often used in environments where these challenges are prevalent (Rao and Mulky 2023). Applying protective coatings to equipment surfaces can help shield against the effects of corrosion and erosion. Cathodic protection systems, which involve the use of sacrificial anodes or impressed currents, can be employed to mitigate corrosion in pipelines and equipment. Routine inspections and monitoring of equipment are essential for detecting and addressing corrosion and erosion before they cause significant damage. The use of chemical inhibitors can help mitigate the effects of corrosion and erosion by reducing the aggressiveness of the fluids and substances in the system.

The consequences of corrosion, erosion, and material degradation in the oil and gas industry are also significant. Corrosion and erosion can significantly reduce the operational lifespan of equipment, necessitating more frequent replacements. Equipment failures due to corrosion and erosion can cause unplanned production disruptions and increased downtime. The structural integrity of equipment can be compromised, posing safety risks to personnel and the environment. Frequent maintenance, repairs, and replacements due to material degradation can lead to increased operational costs.

Thus, the deposition of hydrocarbons can lead to corrosion, erosion, and material degradation, posing serious challenges to the oil and gas industry. Preventive measures, proper materials selection, and regular maintenance are vital to mitigating the effects of these issues and ensuring safe and efficient operations.

13.7 Maintenance and Downtime Considerations

The deposition of hydrocarbons within the oil and gas industry not only affects the efficiency of production equipment but also significantly impacts maintenance requirements and downtime. To manage hydrocarbon deposition and its associated challenges effectively, rigorous monitoring and inspection techniques are imperative. Frequent visual inspections and non-destructive testing are essential for identifying the extent of deposit buildup and the condition of production equipment. The use of advanced technologies such as ultrasonic testing, thermal imaging, and corrosion sensors can provide real-time data on equipment conditions. Risk-based inspection approaches help prioritize maintenance efforts by identifying high-risk areas and equipment that require immediate attention.

Managing the effects of hydrocarbon deposition necessitates a range of cleaning and maintenance techniques. Pigging, scraping, and brushing are employed to physically remove deposits from pipelines and equipment. The use of chemicals, including solvents and dispersants, can dissolve or dislodge deposits. Hot oil can be applied to equipment to melt and remove deposits, particularly in the case of wax buildup. For wellbore blockages, well workovers may be necessary, involving the removal of tubing or equipment for cleaning or replacement. Routine maintenance activities should be scheduled to prevent deposit buildup and address equipment degradation before it becomes severe.

The consequences of maintenance and downtime related to hydrocarbon deposition are multifaceted, such as operational disruption, cost considerations, safety considerations, and environmental concerns. Downtime for maintenance and cleaning operations can disrupt production schedules, leading to missed production targets and potential revenue losses. Maintenance and cleaning operations, especially those conducted under emergency conditions, can be costly. The expenses include labor, equipment, and the associated downtime. Maintenance activities in a complex, high-pressure environment come with safety risks that must be managed carefully. Unplanned downtime can lead to the release of gases and chemicals into the environment, potentially causing environmental damage and regulatory compliance issues.

To mitigate the impact of maintenance and downtime associated with hydrocarbon deposition, practices, such as proactive maintenance, strategic scheduling, training and safety protocols, and environmental responsibility are essential. Implementing a proactive maintenance approach that includes regular inspections, preventive measures, and the use of monitoring technologies is important. Scheduled maintenance during planned shutdowns or during periods of lower production can minimize the impact on overall production targets. Ensuring that personnel involved in maintenance activities are well-trained and establish comprehensive safety protocols to minimize risk. Adhering to environmental regulations and employing practices that minimize the environmental impact of downtime.

Thus, the management of hydrocarbon deposition's maintenance and downtime considerations is crucial for maintaining operational efficiency, safety, and cost-effectiveness in the oil and gas industry. By adopting proactive measures and careful planning, the industry can reduce the impact of hydrocarbon deposition and ensure consistent, reliable production.

13.8 Lost Production Opportunities and Economic Implications

The deposition of hydrocarbons in the oil and gas industry not only leads to operational challenges but also results in lost production opportunities and significant economic implications. In this section, we will examine how these losses occur and the economic consequences they pose to the industry. The timing of production in the oil and gas industry can be critically important due to fluctuating market conditions and pricing. Oil and gas prices are subject to fluctuations in response to global demand, geopolitical factors, and market competition. Delays or disruptions in production can result in missed opportunities to capitalize on favorable market conditions. Efficient production, meeting production targets, and reducing downtime are vital for maximizing profitability. Lost production opportunities can impact revenue and overall economic performance.

The oil and gas industry operates under strict production schedules and contractual obligations. Many companies have contractual agreements to supply specific quantities of oil and gas. Failing to meet these commitments can result in financial penalties and damage to business relationships (Dirani and Ponomarenko 2021). Extended downtime and production delays may lead to a faster depletion of reserves, impacting long-term production sustainability.

The economic implications of lost production opportunities and operational disruptions in the oil and gas industry are also substantial. Unplanned maintenance, emergency cleaning operations, and downtime due to hydrocarbon deposition all come with associated operational costs, including labor, equipment, and materials. Missed production opportunities result in reduced revenues, impacting the financial health and profitability of oil and gas operations. Increased operational costs, revenue losses, and the unpredictability of production can expose companies to financial

Table 13.1 Layers of economic impact (Sousa et al. 2022)

Layer	Nature	Costs to be accounted
Intervention due to wax blockages	Direct costs	Remedial measures
		Repair or removal and replacement actions
		Abandonment and decommissioning
		Permanent loss of production (in case of abandonment)
Deferred production		Flow rate reduction to partial sectional blockage
		Production loss to downtime
Indirect production costs and penalties	Indirect costs	Contract breaches and compensations
		Companies' value decreases due to supply shocks, including penalties in signed contracts; loss of opportunities, market share, reliability in production, and reputation
		Penalties and fees due to accidents
Environmental and Socio-economic costs		Environmental costs
		Socio-economic costs

risk. This risk can affect investment decisions, credit ratings, and borrowing costs. Table 13.1 shows the layers of economic impact on oil and gas industry.

To mitigate the economic implications of lost production opportunities, the oil and gas industry must employ strategic measures. Implementation of proactive approaches to minimize the impact of hydrocarbon deposition. Regular monitoring, preventive maintenance, and cleaning operations can help reduce unplanned downtime. Efficient allocation of resources to prioritize critical equipment and high-risk areas. This helps ensure that maintenance efforts are focused where they are needed most. Development of robust production forecasting models that account for potential downtime due to hydrocarbon deposition, market fluctuations, and other variables. Ensuring that contractual agreements are flexible and account for potential production disruptions, providing some protection against financial penalties.

Thus, the economic implications of lost production opportunities due to hydrocarbon deposition are considerable in the oil and gas industry. Effective management, proactive maintenance, and a keen understanding of market dynamics are essential for reducing the economic impact and ensuring long-term financial sustainability.

13.9 Mitigation Strategies

Addressing the challenges posed by hydrocarbon deposition in the oil and gas industry requires a multi-faceted approach. Mitigation strategies are essential to minimize the impacts of deposition on production, equipment, and economics.

Chemical inhibitors and additives are commonly used to prevent and mitigate hydrocarbon deposition. These can be paraffin inhibitors, asphaltene dispersants, wax dispersants, and scale inhibitors. Paraffin inhibitors are added to crude oil to prevent paraffin wax deposition. They work by modifying the crystal structure of wax, preventing it from adhering to surfaces (Bai et al. 2019). Asphaltene dispersants are used to keep asphaltenes in suspension, preventing their aggregation and deposition. Similar to asphaltene dispersants, these chemicals maintain wax in a dispersed state in the oil, preventing wax buildup (Firoozinia et al. 2016). In addition to hydrocarbon deposits, scales can also form in the production system. Scale inhibitors prevent the formation of scales and their deposition.

Mechanical cleaning techniques are employed to physically remove deposits and prevent their accumulation. Pipeline pigging involves using specialized devices, known as pigs, to scrape the interior surfaces of pipelines. This method is effective in removing deposits and preventing their buildup. Manual or automated scraping and brushing can be used to clean equipment surfaces and remove deposits. Ultrasonic cleaning technology uses high-frequency sound waves to dislodge and remove deposits.

For pipelines, in particular, pigging operations and ongoing maintenance are crucial. The proactive approach such as the implementation of a regular pigging schedule to clean and inspect pipelines can prevent blockages and maintain optimal flow. The use of intelligent pigging tools equipped with sensors to not only clean but also inspect the condition of pipelines can help identify and address potential issues early (Yao et al. 2022). Integrate corrosion management into pigging operations to address corrosion-related issues in addition to deposition. Figure 13.8 shows some mandrel pigs used in pipeline pigging operations.

Fig. 13.8 Some mandrel pigs used in pipeline pigging operations (Guo et al. 2017)

Employing advanced monitoring and surveillance technologies enhances the ability to detect, assess, and address deposition-related challenges. Implement real-time monitoring systems that can detect changes in flow rates, pressure, and equipment conditions. These systems provide early warnings of deposition issues. Utilize digital twin technology to create virtual replicas of production equipment and systems. This enables simulation and prediction of deposition-related challenges. Harness the power of data analytics and machine learning to analyze historical data and predict deposition issues based on trends and patterns.

An integrated approach to deposition management involves a comprehensive strategy that combines chemical, mechanical, and monitoring techniques. Develop deposition models that consider factors such as fluid properties, equipment conditions, and operational variables to predict and manage deposition risks. Implement operational best practices, such as controlling flow rates, temperature, and pressure, to reduce deposition risks. Ensure personnel are well-trained in deposition management, including the safe and effective use of chemicals, equipment, and monitoring systems.

Mitigation strategies should also address sustainability and environmental considerations. Ensure that the use of chemicals and the disposal of waste materials comply with environmental regulations. Aim to reduce the environmental impact of operations by minimizing the release of hydrocarbons and chemicals into the environment. Incorporate sustainable practices, such as using environmentally friendly chemicals, in the deposition management strategy.

Thus, effective mitigation strategies for hydrocarbon deposition in the oil and gas industry encompass a range of chemical, mechanical, and technological approaches. An integrated and proactive approach, considering environmental and sustainability aspects, is essential for minimizing the impacts of deposition and ensuring the reliability and efficiency of production processes.

13.10 Future Perspectives and Innovations

As the oil and gas industry evolves to meet the challenges posed by hydrocarbon deposition, it continues to explore future perspectives and innovative technologies. In this section, we will delve into emerging trends and advancements that are reshaping the industry's approach to deposition management. The oil and gas industry is constantly seeking cutting-edge solutions to address deposition-related issues. Several emerging technologies and research areas hold promise. The application of nanomaterials and nanotechnology is showing potential for preventing deposition at the nanoscale, offering precise and effective solutions (Wang et al. 2022). The development of smart coatings that can actively repel or mitigate deposition is an area of active research. The use of predictive analytics and machine learning to anticipate deposition issues, enabling proactive intervention and prevention. The deployment of robots and drones for remote inspection and cleaning of equipment and pipelines reduces the need for human intervention. The use of advanced sensors and sensor networks for real-time monitoring allows for quicker responses to changing deposition conditions.

The future of deposition management will likely involve a more integrated approach. The development of holistic, interconnected systems that monitor and manage deposition in real-time, optimizing operations and preventing downtime. Greater utilization of digital twin technology to simulate, predict, and manage deposition scenarios in a virtual environment. Integration of artificial intelligence and machine learning into decision support systems for quicker, more informed responses to deposition challenges.

The oil and gas industry is also increasingly focusing on sustainability and environmental responsibility. The development and use of environmentally friendly chemicals for deposition prevention, reducing environmental impact. Exploration of carbon capture and utilization technologies to reduce greenhouse gas emissions associated with oil and gas operations. Continued efforts to improve the energy efficiency of operations, reducing both costs and environmental impact.

The industry is moving towards greater collaboration and knowledge sharing. The development of industry-wide standards and best practices for deposition management to ensure consistency and efficiency. Increased collaboration between industry stakeholders, research institutions, and technology providers to share knowledge and accelerate innovation. Focus on training and education to equip the workforce with the skills needed to implement and manage advanced deposition mitigation technologies.

Thus, the future of deposition management in the oil and gas industry holds significant promise. Emerging technologies, an integrated approach, and a growing emphasis on sustainability and environmental responsibility are poised to transform how the industry addresses deposition-related challenges. Collaboration and knowledge sharing will play a crucial role in implementing these innovations, ensuring the continued efficiency and sustainability of oil and gas operations.

13.11 Conclusion

The deposition of hydrocarbons in the oil and gas industry presents a multifaceted challenge with far-reaching consequences. As we conclude this exploration of the impact, mitigation, and future perspectives related to this issue, it is evident that proactive management of hydrocarbon deposition is critical to the industry's sustainability and success.

In the preceding sections, we have discussed the diverse impacts of hydrocarbon deposition on oil and gas production, encompassing flow restriction, equipment efficiency reduction, plugging, corrosion, erosion, and material degradation. These effects not only hinder operational efficiency but also pose substantial economic risks. Lost production opportunities, maintenance costs, and the financial implications of downtime highlight the need for robust mitigation strategies.

Mitigation strategies, as explored in the content, include the use of chemical inhibitors, mechanical cleaning techniques, advanced monitoring and surveillance technologies, and a holistic, integrated approach to deposition management. These strategies are vital for minimizing the impacts of deposition and ensuring the reliability and efficiency of production processes.

As we look to the future, it's clear that the oil and gas industry is on a path of innovation and sustainability. Emerging technologies, research in nanotechnology, smart coatings, predictive analytics, and the broader use of digital twins, robotics, and advanced sensors are transforming the way the industry addresses deposition challenges. Sustainability and environmental responsibility are taking a prominent role, emphasizing the use of green chemicals, carbon capture, and energy efficiency to reduce the industry's environmental footprint.

Furthermore, collaboration and knowledge sharing are becoming increasingly essential in the industry's approach to deposition management. The development of industry-wide standards and best practices, research collaboration, and workforce training and education are fostering a culture of cooperation and excellence.

In summary, while hydrocarbon deposition in the oil and gas industry is a persistent challenge, it is also a catalyst for innovation, growth, and the development of more sustainable and responsible practices. By embracing emerging technologies, implementing proactive strategies, and emphasizing environmental responsibility, the industry is poised to overcome the challenges of deposition and forge a path toward a more efficient, economically viable, and environmentally conscious future. Through collective efforts and a commitment to excellence, the oil and gas industry is well-equipped to navigate the complexities of hydrocarbon deposition and ensure the long-term success of this critical sector.

References

Alnaimat F, Ziauddin M (2020) Wax deposition and prediction in petroleum pipelines. J Pet Sci Eng 184:106385. https://doi.org/10.1016/j.petrol.2019.106385

Askari M, Aliofkhazraei M, Jafari R, Hamghalam P, Hajizadeh A (2021) Downhole corrosion inhibitors for oil and gas production—a review. Appl Surf Sci Adv 6:100128. https://doi.org/10.1016/j.apsadv.2021.100128

Bai J, Jin X, Wu J-T (2019) Multifunctional anti-wax coatings for paraffin control in oil pipelines. Pet Sci 16:619–631. https://doi.org/10.1007/s12182-019-0309-7

Banki R, Hoteit H, Firoozabadi A (2008) Mathematical formulation and numerical modeling of wax deposition in pipelines from enthalpy–porosity approach and irreversible thermodynamics. Int J Heat Mass Transf 51:3387–3398. https://doi.org/10.1016/j.ijheatmasstransfer.2007.11.012

Burger ED, Perkins TK, Striegler JH (1981) Studies of wax deposition in the Trans Alaska pipeline. JPT J Pet Technol 33:1075–1086. https://doi.org/10.2118/8788-PA

Chala GT, Sulaiman SA, Japper-Jaafar A (2018) Flow start-up and transportation of waxy crude oil in pipelines-A review. J Nonnewton Fluid Mech 251:69–87. https://doi.org/10.1016/j.jnnfm.2017.11.008

Dirani F, Ponomarenko T (2021) Contractual systems in the oil and gas sector: current status and development. Energies 14:5497. https://doi.org/10.3390/en14175497

Fakher S, Ahdaya M, Elturki M, Imqam A (2020) Critical review of asphaltene properties and factors impacting its stability in crude oil. J Pet Explor Prod Technol 10:1183–1200. https://doi.org/10.1007/s13202-019-00811-5

Firoozinia H, Fouladi Hossein Abad K, Varamesh A (2016) A comprehensive experimental evaluation of asphaltene dispersants for injection under reservoir conditions. Pet Sci 13:280–291. https://doi.org/10.1007/s12182-016-0078-5

Guo B, Liu X, Tan X (2017) Pipeline pigging. Pet Prod Eng 701–720. https://doi.org/10.1016/B978-0-12-809374-0.00022-2

Han S, Huang Z, Senra M, Hoffmann R, Fogler HS (2010) Method to determine the wax solubility curve in crude oil from centrifugation and high temperature gas chromatography measurements. Energy Fuels 24:1753–1761. https://doi.org/10.1021/ef901195w

Hansen AH (2023) Flow through restriction, 43–61. https://doi.org/10.1007/978-3-031-15089-0_4

Li N, Mao GL, Shi XZ, Tian SW, Liu Y (2018) Advances in the research of polymeric pour point depressant for waxy crude oil. J Dispers Sci Technol 39:1165–1171. https://doi.org/10.1080/01932691.2017.1385484

Mohamed NH, Soliman FS, El Maghraby H, Moustfa YM (2017) Thermal conductivity enhancement of treated petroleum waxes, as phase change material, by α nano alumina: energy storage. Renew Sustain Energy Rev 70:1052–1058. https://doi.org/10.1016/j.rser.2016.12.009

Oon CS, Kazi SN, Hakimin MA, Abdelrazek AH, Mallah AR, Low FW, Tiong SK, Badruddin IA, Kamanger S (2020) Heat transfer and fouling deposition investigation on the titanium coated heat exchanger surface. Powder Technol 373:671–680. https://doi.org/10.1016/j.powtec.2020.07.010

Rao P, Mulky L (2023) Erosion-corrosion of materials in industrial equipment: a review. ChemElectroChem 10. https://doi.org/10.1002/celc.202300152

Raya SA, Mohd Saaid I, Abbas Ahmed A, Abubakar Umar A (2020) A critical review of development and demulsification mechanisms of crude oil emulsion in the petroleum industry. J Pet Explor Prod Technol 10:1711–1728. https://doi.org/10.1007/s13202-020-00830-7

Singhai HK, Sahai GC, Pundeer GS, Chandra K (1991) Designing and selecting wax crystal modifier for optimum field performance based on crude oil composition. https://doi.org/10.2118/22784-ms

Sousa AL, Matos HA, Guerreiro LP (2019) Preventing and removing wax deposition inside vertical wells: a review. J Pet Explor Prod Technol 9:2091–2107. https://doi.org/10.1007/s13202-019-0609-x

Sousa AM, Ribeiro TP, Pereira MJ, Matos HA (2022) On the economic impact of wax deposition on the oil and gas industry. Energy Convers Manag X 16:100291. https://doi.org/10.1016/j.ecmx.2022.100291

Speight JG (2014) The chemistry and technology of petroleum. Chem Technol Pet. https://doi.org/10.1201/b16559

Speight JG (2015) Occurrence and formation of crude oil and natural gas. In: Subsea deep oil and gas science technology. Elsevier, pp 1–43. https://doi.org/10.1016/B978-1-85617-558-6.00001-5

Thanh NX, Hsieh M, Philp RP (1999) Waxes and asphaltenes in crude oils. Org Geochem 30:119–132. https://doi.org/10.1016/S0146-6380(98)00208-3

Theyab MA, Diaz PA (2016) Experimental study on the effect of spiral flow on wax deposition volume. SPE. https://doi.org/10.2118/182936-MS

Wang C, Chen H, Shi H, Ma K, Ma Q, Gong J (2022) Role of a nanocomposite pour point depressant on wax deposition in different flow patterns from the perspective of crystallization kinetics. ACS Omega 7:11200–11207. https://doi.org/10.1021/acsomega.2c00068

Yang F, Li C, Li C, Wang D (2013) Scaling of structural characteristics of gelled model waxy oils. Energy Fuels 27:3718–3724. https://doi.org/10.1021/ef400554v

Yao B, He Z, Lu N, Zhang S (2022) A novel PIG and an intelligent pigging scheme based on deep-learning technology. Int J Press Vessel Pip 200:104803. https://doi.org/10.1016/j.ijpvp.2022.104803

Zuo P, Qu S, Shen W (2019) Asphaltenes: Separations, structural analysis and applications. J Energy Chem 34:186–207. https://doi.org/10.1016/j.jechem.2018.10.004

The manufacturer's authorised representative in the EU is Springer Nature Customer Service Centre GmbH, Europaplatz 3, 69115 Heidelberg, Germany. If you have any concerns regarding our products, please contact ProductSafety@springernature.com

Printed and bound by CPI Group (UK) Ltd, Croydon, CR0 4YY
30/09/2026
02225596-0002